Springer Tracts in Modern Physics

Volume 266

Springer Tracts in Modern Physics

Springer Tracts in Modern Physics provides comprehensive and critical reviews of topics of current interest in physics. The following fields are emphasized: Elementary Particle Physics, Condensed Matter Physics, Light Matter Interaction, Atomic and Molecular Physics, Complex Systems, Fundamental Astrophysics.

Suitable reviews of other fields can also be accepted. The editors encourage prospective authors to correspond with them in advance of submitting a manuscript. For reviews of topics belonging to the above mentioned fields, they should address the responsible editor as listed below.

Special offer: For all clients with a print standing order we offer free access to the electronic volumes of the Series published in the current year.

More information about this series at http://www.springer.com/series/426

Xianping Wang · Cheng Yin
Zhuangqi Cao

Progress in Planar Optical Waveguides

上海交通大學出版社
SHANGHAI JIAO TONG UNIVERSITY PRESS

Springer

Xianping Wang
College of Physics and Communication Electronics
Jiangxi Normal University
Nanchang, Jiangxi
China

Zhuangqi Cao
Shanghai Jiao Tong University
Shanghai
China

Cheng Yin
Hohai University
Changzhou, Jiangsu
China

ISSN 0081-3869 ISSN 1615-0430 (electronic)
Springer Tracts in Modern Physics
ISBN 978-3-662-48982-6 ISBN 978-3-662-48984-0 (eBook)
DOI 10.1007/978-3-662-48984-0

Jointly published with Shanghai Jiao Tong University Press

Library of Congress Control Number: 2015955389

Springer Heidelberg New York Dordrecht London

Printed on acid-free paper

Springer-Verlag GmbH Berlin Heidelberg is part of Springer Science+Business Media
(www.springer.com)

Preface

Since the 1960s, the waveguide optics has evolved into an emerging discipline and had a tremendous impact on our information society. It attracted considerable attention of scientists and technologists because of the potential applications for signal processing, biochemical sensing, etc. The parallel development of material science, optoelectronics, and micromachining technology has overcome several technical obstacles and made possible the implementation of optical waveguide system to perform those fascinating applications. In addition to this, optical waveguide devices with new principles, new materials, and new structures are constantly proposed, and some of them, including planar optical integrated circuits, photonic crystal structure, nano-array structure, and plasmonics, become the research hot spots in academic and industry. In one word, waveguide optics is an extremely promising and fast-growing field.

This book is intended to serve as a general text on planar optical waveguide structure for senior undergraduates. The Chinese version has published 7 years ago, and the idea of writing this book is a result of frequent enquiries about the possibility of publishing an English version. Because many significant advances on waveguide optics have been made during the intervening years, it is felt that a direct translation is hardly appropriate. Instead, a substantially new book is prepared, which I am now placing before the readers. This book is intended to summarize our recent research results and introduce the current progresses in optical waveguides. Therefore, some relatively old contents written in the Chinese version have been omitted. The following characteristics can be found in this book:

1. Some mature theories, such as optical fiber theory, coupled modes theory, and a number of numerical methods, have already analyzed in other equivalent books. These contents are not contained in this book because I designed to restrict its scope to a narrow field, which is only a collection of some of our researches on waveguide optics.

2. Both the analytical transfer matrix methods and the perturbation analysis are the mathematical basis of this book. By applying these two methods, several waveguide optics problems can be depicted with the formulae in the analytical form and the corresponding derivation processes are quite simple.
3. The field distribution function of the optical waveguide is not expressed by the conventional sine or cosine form, but the exponential form, and thus, a transmission-type dispersion equation with more clear physical insight is obtained. Moreover, a newly defined concept, i.e., the scattered subwaves, is firstly proposed. The neglect of the scattered subwaves in other semiclassical theories, such as WKB and SWKB, results in various confusions and paradoxes. Based on this new concept, an analytical dispersion equation for the graded-index waveguide is also given.
4. The Goos–Hänchen shift occurred at the total reflection is an interesting and important issue in waveguide optics. Several causality paradoxes originated from this issue have been discussed, and our debate to these paradoxes is presented in this book. The experimental applications of Goos–Hänchen shift and other non-specular effects are also added.
5. The surface plasmon wave, the metal-cladding waveguide, and the corresponding principle of attenuated total reflection are one of my top and main research topics. In this book, the basic theories and experimental applications of our proposed symmetrical metal-cladding waveguide, free-space coupling technology, ultrahigh-order modes, wideband slow light, conical reflection, and oscillating wave sensors are included.

I would like to express my thanks to my Ph.D. students, Yi Jiang, Xiangming Liu, Fan Chen, Yi Wang, Wen Yuan, Cheng Yin, and Xianping Wang. I would like to express my sincere gratitude to Prof. J.Z. Li (Shenzhen University) and Prof. H. Ming (University of Science and Technology of China) for their help in recommending and editing the English version of the book.

Although great care has been exercised, some errors may still occur in this book. Therefore, I appreciate any comments from readers.

Zhuangqi Cao

Contents

Chapter 1
Basic Analysis on Optical Waveguides

Abstract Dielectric optical waveguide, including circular section waveguide (optical fiber), slab waveguide, and strip waveguide, is used to confine the propagation of light in the application of optical waveguide devices and optical integrated circuits. The working frequency is in the range of visible and near-infrared region. The theory of dielectric optical waveguide is based on the conventional electromagnetic theory and the optical properties of optical materials. In this chapter, the Maxwell's equations governing the optical propagation characteristics of the optical waveguide are introduced and the derived wave equations are given.

Keywords Maxwell's equation · Snell's laws · Fresnel formula · Goos–Hänchen shift

1.1 Basic Theory of Wave Optics

This section briefly summarizes the Maxwell's equations, matter equations, and boundary conditions that are necessary to understand waveguide optics. The important concept such as reflection, transmission, and Goos–Hänchen shift is also introduced.

1.1.1 Maxwell's Equation

Light is a kind of electromagnetic waves, which are related to the four field quantities described by the Maxwell's equations. The electric and magnetic field intensities **E** and **H** are measured in units of [volt/m] and [ampere/m], respectively. The quantities **D** and **B** are referred as the electric and magnetic flux densities and

X. Wang et al., *Progress in Planar Optical Waveguides*,
Springer Tracts in Modern Physics 266, DOI 10.1007/978-3-662-48984-0_1

are in units of [coulomb/m^2] and [weber/m^2], or [tesla]. These four quantities fulfill the following equations:

$$\nabla \times \mathbf{E} = -\frac{\partial \mathbf{B}}{\partial t}, \tag{1.1}$$

$$\nabla \times \mathbf{H} = \frac{\partial \mathbf{D}}{\partial t} + \mathbf{J}, \tag{1.2}$$

where $\mathbf{J}$[A/m^2] is the electric current density, which is related to the volume charge ρ[C/m^3] via the charge conservation or current continuity equation

$$\nabla \cdot \mathbf{J} = -\frac{\partial \rho}{\partial t}, \tag{1.3}$$

in view of the identical equation $\nabla \cdot (\nabla \times \mathbf{A}) = 0$. It is easy to obtain the following equations from Eqs. (1.1) and (1.2):

$$\nabla \cdot \mathbf{B} = 0, \tag{1.4}$$

$$\nabla \cdot \mathbf{D} = \rho, \tag{1.5}$$

The above Eqs. (1.1), (1.2), (1.4), and (1.5) form the Maxwell's equations.

1.1.2 Matter Equation

In order to determine all the field vector functions from the given current and charge distribution, equations on the material electromagnetic properties should be added. These equations are called as the matter equations, or the constitutive relations. For the linear, isotropic, and non-magnetic media, they can be written as

$$\mathbf{D} = \varepsilon \mathbf{E}, \tag{1.6}$$

$$\mathbf{B} = \mu \mathbf{H}, \tag{1.7}$$

where the permittivity ε and the permeability μ of a material are defined by

$$\varepsilon = \varepsilon_0 \varepsilon_r, \tag{1.8}$$

$$\mu = \mu_0 \mu_r, \tag{1.9}$$

where ε_0 and μ_0 refer to the permittivity and permeability of vacuum, respectively, and ε_r and μ_r are known as the relative permittivity and permeability, respectively. For a non-magnetic material, we have $\mu_r = 1$. If the light speed in vacuum is c, then there will be

$$\varepsilon_0 = \frac{1}{c^2 \mu_0} \approx 8.854188 \times 10^{-12} \text{ [F/m]}, \tag{1.10}$$

$$\mu_0 = 4\pi \times 10^{-7} \text{[H/m]}. \tag{1.11}$$

In the conductor, the relation between the current density **J** and the electric field **E** is defined via the conductivity σ[S/m]

$$\mathbf{J} = \sigma \mathbf{E}. \tag{1.12}$$

The material's refractive index is defined by

$$n = \sqrt{\varepsilon_r}. \tag{1.13}$$

In isotropic materials, ε and μ are scalar quantities, but for anisotropic materials, these quantities depend on the x, y, and z directions and should be written as tensors. Consequently, Eqs. (1.6) and (1.7) should be transferred into

$$D_i = \sum_{j=1}^{3} \varepsilon_{ij} E_j, \tag{1.14}$$

$$B_i = \sum_{j=1}^{3} \mu_{ij} H_j, \tag{1.15}$$

where $i, j = 1, 2, 3$ that are corresponding to the x, y, and z directions of the Cartesian coordinate. If the media is non-absorptive, we have

$$\varepsilon_{ji}^{*} = \varepsilon_{ij}, \tag{1.16}$$

where the superscript "*" represents the conjugate of a complex number.

1.1.3 Wave Equation

By imagining a plane wave of angular frequency ω[rad/s] propagating in a homogeneous lossless medium, and by assuming the harmonic time dependence, we can cast its field components into

$$\mathbf{E}(\mathbf{r}, t) = \text{Re}[\mathbf{E}(\mathbf{r})\exp - (i\omega t)], \tag{1.17}$$

$$\mathbf{H}(\mathbf{r}, t) = \text{Re}[\mathbf{H}(\mathbf{r})\exp(-i\omega t)], \tag{1.18}$$

$$\mathbf{D}(\mathbf{r}, t) = \text{Re}[\mathbf{D}(\mathbf{r})\exp(-i\omega t)], \tag{1.19}$$

$$\mathbf{B}(\mathbf{r},t) = \mathrm{Re}[\mathbf{B}(\mathbf{r})\ \exp(-i\omega t)]. \tag{1.20}$$

In the absence of free charge and current densities ($\rho = 0, \mathbf{J} = 0$), and assuming non-magnetic ($\mu_r = 1$), the time-independent Maxwell's equation can be written as

$$\nabla \times \mathbf{E} = i\omega\mathbf{B} = i\omega\mu_0\mathbf{H}, \tag{1.21}$$

$$\nabla \times \mathbf{H} = -i\omega\mathbf{D} = -i\omega\varepsilon\mathbf{E}, \tag{1.22}$$

$$\nabla \cdot \mathbf{H} = 0, \tag{1.23}$$

$$\nabla \cdot (\varepsilon_r\mathbf{E}) = 0. \tag{1.24}$$

Applying the curl to Eq. (1.21), and substituting it into Eq. (1.22), one will have

$$\nabla \times (\nabla \times \mathbf{E}) = \omega^2\varepsilon_0\mu_0\varepsilon_r\mathbf{E}. \tag{1.25}$$

In view of

$$\nabla \times (\nabla \times \mathbf{A}) = \nabla(\nabla \cdot \mathbf{A}) - \nabla^2\mathbf{A}, \tag{1.26}$$

one can recast Eq. (1.24) into

$$\nabla \cdot (\varepsilon_r\mathbf{E}) = \nabla\varepsilon_r \cdot \mathbf{E} + \varepsilon_r\nabla \cdot \mathbf{E} = 0,$$

and can obtain

$$\nabla \cdot \mathbf{E} = -\frac{\nabla\varepsilon_r}{\varepsilon_r} \cdot \mathbf{E}, \tag{1.27}$$

Combining Eqs. (1.26) and (1.27), the vector wave equation for the electric field **E** takes the form

$$\nabla^2\mathbf{E} + \nabla\left(\frac{\nabla\varepsilon_r}{\varepsilon_r} \cdot \mathbf{E}\right) + k_0^2\varepsilon_r\mathbf{E} = 0, \tag{1.28}$$

where the wave number k_0 of the vacuum is given by

$$k_0 = \omega\sqrt{\varepsilon_0\mu_0} = \frac{\omega}{c}, \tag{1.29}$$

and the wave number k in the material can be obtained by the refractive index $n = \sqrt{\varepsilon_r}$ and written as

$$k = k_0 n. \tag{1.30}$$

If the relative permittivity of the material is constant, then we can simplify the vector wave Eq. (1.28) into the Helmholtz equation

$$\nabla^2 \mathbf{E} + k^2 \mathbf{E} = 0. \tag{1.31}$$

Using a similar procedure, it is not difficult to derive the Helmholtz equation for magnetic field **H**

$$\nabla^2 \mathbf{H} + k^2 \mathbf{H} = 0. \tag{1.32}$$

1.1.4 Boundary Condition for Electromagnetic Field

Consider an interface between two materials with a different refractive index as plotted in Fig. 1.1, and let **n** be the unit vector in the normal direction of the interface; then, the boundary condition can be expressed as

(1) The components of **E**-field parallel to the interface are continuous across the boundary

$$\mathbf{n} \times (\mathbf{E}_1 - \mathbf{E}_2) = 0, \tag{1.33}$$

or

$$E_{1t} = E_{2t}, \tag{1.34}$$

where t denotes the tangential direction.

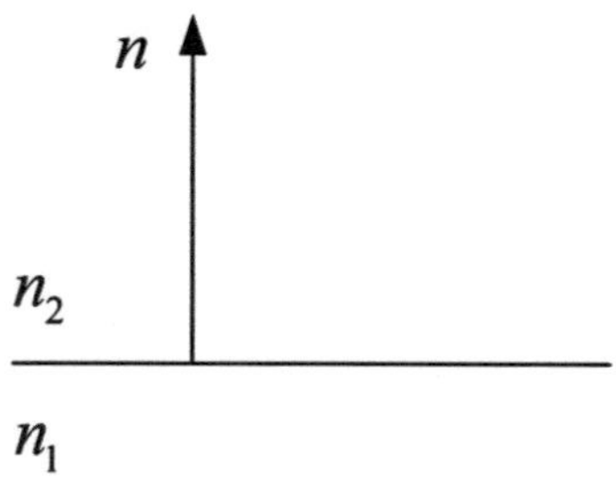

Fig. 1.1 Interface between two dielectrics with refractive index of n_1 and n_2

(2) The parallel components of **H**-field are discontinuous across the boundary due to the existence of surface current density.

$$\mathbf{n} \times (\mathbf{H}_1 - \mathbf{H}_2) = \mathbf{J}_S, \tag{1.35}$$

where $\mathbf{J}_S$ [A/m] represents the surface current density, in case that $\mathbf{J}_S = 0$, there is

$$\mathbf{n} \times (\mathbf{H}_1 - \mathbf{H}_2) = 0. \tag{1.36}$$

So if there is no surface current, the tangential component of the magnetic field is continuous across the boundary, and Eq. (1.36) is equivalent to

$$H_{1t} = H_{2t}. \tag{1.37}$$

(3) The component of **D**-field perpendicular to the interface is discontinuous if the surface charge density is nonzero

$$\mathbf{n} \cdot (\mathbf{D}_1 - \mathbf{D}_2) = \rho_S, \tag{1.38}$$

where ρ_S [C/m^2] is the surface charge density. In case of the absence of surface charge, there is

$$\mathbf{n} \cdot (\mathbf{D}_1 - \mathbf{D}_2) = 0, \tag{1.39}$$

or equivalently

$$D_{1n} = D_{2n}, \tag{1.40}$$

which can be explained as the normal component of the **D** that is continuous at the boundary.

(4) The normal component of *B*-field at the boundary is continuous

$$\mathbf{n} \cdot (\mathbf{B}_1 - \mathbf{B}_2) = 0, \tag{1.41}$$

or equivalently

$$B_{1n} = B_{2n}. \tag{1.42}$$

1.1.5 Poynting Vector

Because the electric potential **U** is the integral of the electric field, and magnetic field can produce current, as a result it is natural to relate the cross product of the electric field and the magnetic field with the energy of the electromagnetic field. If the operator ∇ is applied to $\mathbf{E}\times\mathbf{H}$, it follows

$$\nabla\cdot(\mathbf{E}\times\mathbf{H})=\mathbf{H}\cdot(\nabla\times\mathbf{E})-\mathbf{E}\cdot(\nabla\times\mathbf{H}). \tag{1.43}$$

Substituting Eqs. (1.1), (1.2), and (1.12) into Eq. (1.43), there will be

$$\begin{aligned}\nabla\cdot(\mathbf{E}\times\mathbf{H})&=-\mu\mathbf{H}\cdot\frac{\partial\mathbf{H}}{\partial t}-\varepsilon\mathbf{E}\cdot\frac{\partial\mathbf{E}}{\partial t}-\sigma\mathbf{E}^2\\&=-\frac{\partial}{\partial t}\left(\frac{1}{2}\varepsilon\mathbf{E}^2+\frac{1}{2}\mu\mathbf{H}^2\right)-\sigma\mathbf{E}^2.\end{aligned} \tag{1.44}$$

Taking the integral of (1.44) over a volume in a closed surface, and applying the Gauss theorem, we will have

$$\begin{aligned}\int_V\nabla\cdot(\mathbf{E}\times\mathbf{H})\mathrm{d}V&=\int_S(\mathbf{E}\times\mathbf{H})\cdot\mathbf{n}\mathrm{d}S\\&=-\frac{\partial}{\partial t}\int_V\left(\frac{1}{2}\varepsilon\mathbf{E}^2+\frac{1}{2}\mu\mathbf{H}^2\right)\mathrm{d}V-\int_V\sigma\mathbf{E}^2\mathrm{d}V,\end{aligned} \tag{1.45}$$

where S is the closed surface, V is the volume inside, and $\mathbf{n}$ denotes the unit vector of the normal direction of the surface. The first integral on the right-hand side is the total energy stored in both the electric field and the magnetic field. The second term represents the work done (or the Joule heat) during per unit time. Evidently, the term $\int_S(\mathbf{E}\times\mathbf{H})\cdot\mathbf{n}\mathrm{d}S$ denotes the rate at which energy is carried out through the closed surface.

The Poynting vector

$$\mathbf{S}=\mathbf{E}\times\mathbf{H}\ [\mathrm{w/m^2}], \tag{1.46}$$

is the energy transported by the field per unit time and per unit area, and it is also known as the energy flux density. If we replace the outward pointing vector **n** by the inward pointing vector $\boldsymbol{\mu}$, it is possible to express the power entering into the volume V though the surface by

$$P=\int_S-(\mathbf{E}\times\mathbf{H})\cdot\mathbf{n}\mathrm{d}S=\int_S(\mathbf{E}\times\mathbf{H})\cdot\boldsymbol{\mu}\mathrm{d}S. \tag{1.47}$$

According to Eqs. (1.17) and (1.18), the electromagnetic field can be written in the complex amplitude form

$$\mathbf{E}(\mathbf{r},t) = \frac{1}{2}[\mathbf{E}(\mathbf{r})\mathrm{e}^{-\mathrm{i}\omega t} + \mathbf{E}^*(\mathbf{r})\mathrm{e}^{\mathrm{i}\omega t}], \tag{1.48}$$

$$\mathbf{H}(\mathbf{r},t) = \frac{1}{2}[\mathbf{H}(\mathbf{r})\mathrm{e}^{-i\omega t} + \mathbf{H}^*(\mathbf{r})\mathrm{e}^{i\omega t}]. \tag{1.49}$$

Consequently, the time-averaged normal component of the Poynting vector is

$$\begin{aligned} \langle \mathbf{S} \cdot \boldsymbol{\mu} \rangle &= \langle (\mathbf{E} \times \mathbf{H}) \cdot \boldsymbol{\mu} \rangle \\ &= \frac{1}{4}\langle [(\mathbf{E}\mathrm{e}^{-i\omega t} + \mathbf{E}^*\mathrm{e}^{i\omega t}) \times (\mathbf{H}\mathrm{e}^{-i\omega t} + \mathbf{H}^*\mathrm{e}^{i\omega t})] \cdot \boldsymbol{\mu} \rangle \\ &= \frac{1}{4}\langle [\mathbf{E} \times \mathbf{H}^* + \mathbf{E}^* \times \mathbf{H} + \mathbf{E} \times \mathbf{H}\mathrm{e}^{-i2\omega t} + \mathbf{E}^* \times \mathbf{H}^*\mathrm{e}^{i2\omega t}] \cdot \boldsymbol{\mu} \rangle \\ &= \frac{1}{4}(\mathbf{E} \times \mathbf{H}^* + \mathbf{E}^* \times \mathbf{H}) \cdot \boldsymbol{\mu} \\ &= \frac{1}{2}\mathrm{Re}[(\mathbf{E} \times \mathbf{H}^*) \cdot \boldsymbol{\mu}] \end{aligned} \tag{1.50}$$

where $< >$ represents the time average. The time-averaged power flow can be written as

$$P = \int_S \frac{1}{2}\mathrm{Re}[(\mathbf{E} \times \mathbf{H}^*) \cdot \boldsymbol{\mu}]\mathrm{d}S. \tag{1.51}$$

In the analysis of optical waveguide, $\mathbf{E} \times \mathbf{H}^*$ is usually real, so we can also directly write Eq. (1.51) as

$$P = \int_S \frac{1}{2}(\mathbf{E} \times \mathbf{H}^*) \cdot \boldsymbol{\mu}\mathrm{d}S. \tag{1.52}$$

1.2 Reflection and Transmission of the Plane Waves

This section summarizes the reflection and transmission of an incident plane wave at the interface between two different dielectric media without presenting the derivation.

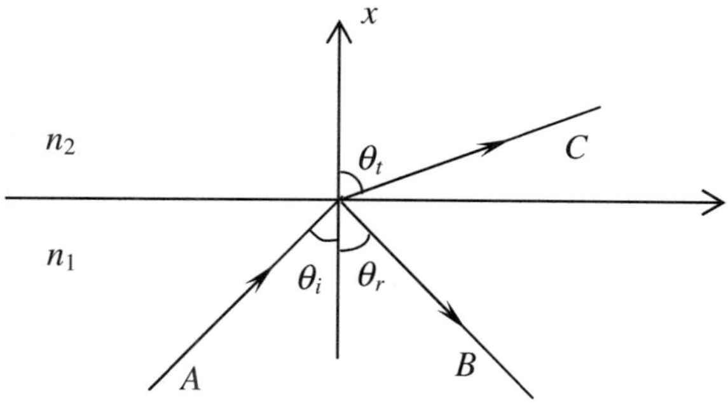

Fig. 1.2 Reflection and transmission at a planar interface separating two media

1.2.1 *Snell's Laws*

Consider a planar interface at $x = 0$, which separates two lossless isotropic media n_1 and n_2 ($n_1 > n_2$). In Fig. 1.2, alphabet "A" denotes the complex amplitude of a monochromatic plane wave incident on medium n_1, which results in a reflected wave with a complex amplitude "B" in the same media and a transmitted wave with a complex amplitude "C" in the second media. Assuming that the incident, reflected, and transmitted angles are $\theta_{\rm i}$, $\theta_{\rm r}$, and $\theta_{\rm t}$, respectively, then it is simple to draw the following conclusions based on the boundary conditions:

(1) Angles of reflection and incidence are equal

$$\theta_{\rm i} = \theta_{\rm r}, \tag{1.53}$$

which is known as Snell's law of reflection.

(2) The transmitted angle $\theta_{\rm t}$ and that of incidence $\theta_{\rm i}$ obey the Snell's law of refraction, which has the form

$$n_1 \sin\theta_{\rm i} = n_2 \sin\theta_{\rm t}. \tag{1.54}$$

1.2.2 *Fresnel Formula*

While the relations between the angles of incident, reflected, and transmitted waves are specified by the Snell's law, the Fresnel formula determined the relation of the amplitude and phase of the three waves. Usually, the Fresnel reflection coefficient determines the ratio of the complex amplitude B of the reflected wave and the incident wave, and the Fresnel transmission coefficient gives the ratio of the

Fig. 1.3 TE-polarized light incident on an interface ($n_1 > n_2$)

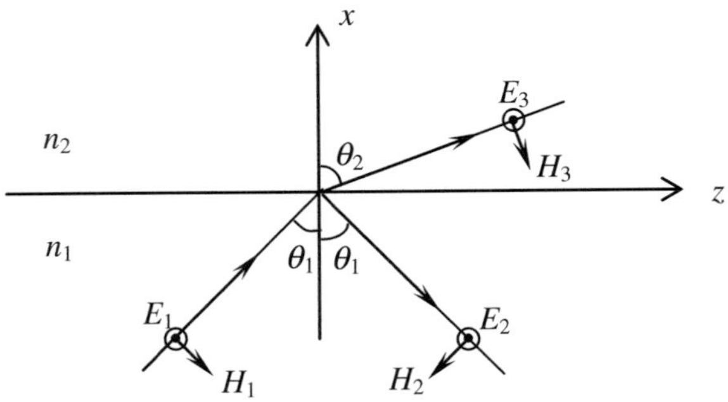

complex amplitude C of the transmitted wave and the incident wave. These coefficients depend on the incident angle and the polarization of the light.

(1) Figure 1.3 shows a TE-polarized incident wave, whose electric field is perpendicular to the plane of incidence, and its magnetic fields lie in the plane. For the TE polarization, we can obtain

(1.1) The reflection coefficient is

$$r_{\mathrm{TE}} = \frac{n_1 \cos\theta_1 - n_2 \cos\theta_2}{n_1 \cos\theta_1 + n_2 \cos\theta_2} = \frac{\sin(\theta_2 - \theta_1)}{\sin(\theta_2 + \theta_1)}. \tag{1.55}$$

(1.2) The transmission coefficient is

$$t_{\mathrm{TE}} = \frac{2n_1 \cos\theta_1}{n_1 \cos\theta_1 + n_2 \cos\theta_2} = \frac{2\sin\theta_2 \cos\theta_1}{\sin(\theta_2 + \theta_1)}. \tag{1.56}$$

(2) Figure 1.4 shows a TM-polarized incident wave, whose magnetic field is perpendicular to the plane of incidence, and its electric fields lie in the plane. For the TM polarization, we can obtain

Fig. 1.4 TM-polarized light incident on an interface ($n_1 > n_2$)

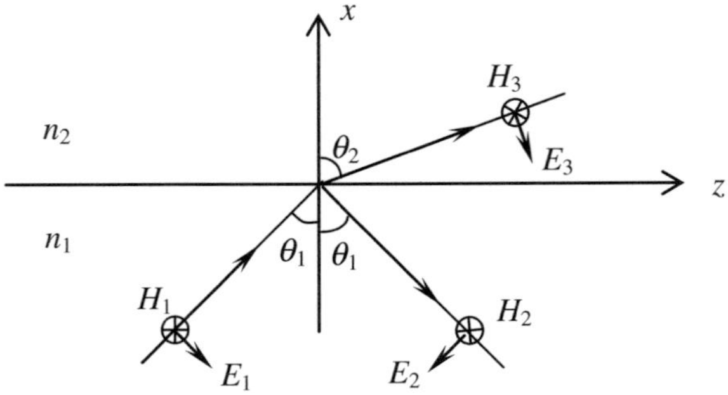

(2.1) The reflection coefficient is

$$r_{\mathrm{TM}} = \frac{n_2 \cos\theta_1 - n_1 \cos\theta_2}{n_2 \cos\theta_1 + n_1 \cos\theta_2} = \frac{\tan(\theta_1 - \theta_2)}{\tan(\theta_1 + \theta_2)}. \tag{1.57}$$

(2.2) The transmission coefficient is

$$t_{\mathrm{TM}} = \frac{2n_2 \cos\theta_1}{n_2 \cos\theta_1 + n_1 \cos\theta_2} = \frac{\sin 2\theta_1}{\sin(\theta_1 + \theta_2)\cos(\theta_1 - \theta_2)}. \tag{1.58}$$

For both polarizations, their reflection probability can be defined as

$$R = rr^* = |r|^2, \tag{1.59}$$

and their transmission probability is given by

$$T = tt^* = |t|^2. \tag{1.60}$$

It is easy to demonstrate that for both the TE and TM polarizations, there is

$$R + T = 1, \tag{1.61}$$

which obeys the conservation of energy.

1.2.3 Brewster's Angle

According to the Snell's law of refraction, when the condition

$$\theta_1 + \theta_2 = \frac{\pi}{2}, \tag{1.62}$$

is fulfilled, there is $\tan(\theta_1 + \theta_2) = \infty$, so that $R_{\mathrm{TM}} = 0$ and $R_{\mathrm{TE}} \neq 0$. As is shown in Fig. 1.5, if both TM and TE waves are incident on the interface, only TE-polarized waves will be reflected in the case that the reflected light and the transmitted light are perpendicular to each other.

Using $n_1 \sin\theta_1 = n_2 \sin\theta_2$ and Eq. (1.62), we find

$$\sin\theta_2 = \sin\left(\frac{\pi}{2} - \theta_1\right) = \cos\theta_1.$$

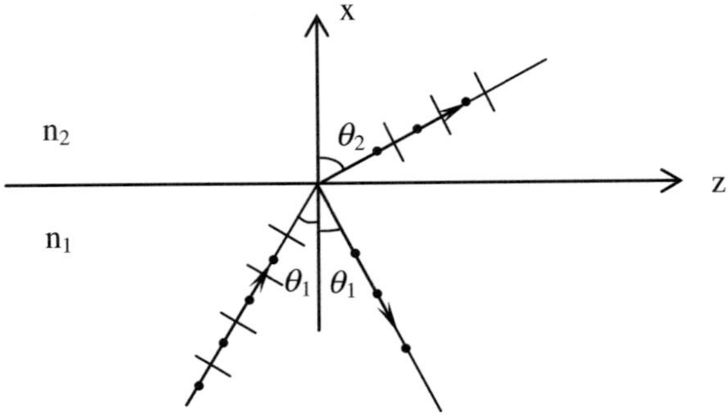

Fig. 1.5 Brewster's angle

Together with the Snell's law of refraction, one can deduce that

$$\tan\theta_1 = \frac{n_2}{n_1}.$$

So the Brewster's angle is found to be

$$\theta_B = \tan^{-1}\left(\frac{n_2}{n_1}\right). \tag{1.63}$$

In principle, it is possible to generate polarized waves via this Brewster's angle, but the efficiency is very low, since the reflection probability of the TE mode is also small. The widely used method to generate a polarization wave is based on the multilayer structures.

1.2.4 Total Internal Reflection

From Eqs. (1.55) and (1.57), it is clear that if the light is incident from a denser to a lighter material, the transmission angle can reach its maximum value

$$\theta_2 = \frac{\pi}{2},$$

which follows

$$r_{TE} = r_{TM} = 1,$$

and

$$t_{TE} = t_{TM} = 0.$$

Meanwhile, the incident angle θ_1 can be calculated via $n_1 \sin\theta_1 = n_2$, which is known as the critical angle of incidence. We can rewrite this angle in the form

$$\theta_c = \sin^{-1}\frac{n_2}{n_1}. \tag{1.64}$$

In summary, when a laser is incident from a denser to a lighter dielectric with incident angle equal to the critical angle $\theta_1 = \theta_c$, the transmitted laser travels along the interface; when the incident angle is larger than the critical angle $\theta_1 > \theta_c$, all the energy returns to the first dielectric, and this phenomenon is known as the total internal reflection. Since there is $n_1 \sin\theta_1 > n_2$, the reflection coefficient is no longer real, indicating that the reflected laser suffers a phase shift which is dependent on its polarization.

(1) TE Polarization

From Eq. (1.55), there is

$$\begin{aligned} r_{TE} &= \frac{n_1\cos\theta_1 - n_2\cos\theta_2}{n_1\cos\theta_1 + n_2\cos\theta_2} \\ &= \frac{\sqrt{n_1^2 - n_1^2\sin^2\theta_1} - \sqrt{n_2^2 - n_1^2\sin^2\theta_1}}{\sqrt{n_1^2 - n_1^2\sin^2\theta_1} + \sqrt{n_2^2 - n_1^2\sin^2\theta_1}} \\ &= \frac{\sqrt{n_1^2 - n_1^2\sin^2\theta_1} - i\sqrt{n_1^2\sin^2\theta_1 - n_2^2}}{\sqrt{n_1^2 - n_1^2\sin^2\theta_1} + i\sqrt{n_1^2\sin^2\theta_1 - n_2^2}} \\ &= \exp(-i2\Phi_{TE}), \end{aligned} \tag{1.65}$$

where

$$\Phi_{TE} = \tan^{-1}\left(\sqrt{\frac{n_1^2\sin^2\theta_1 - n_2^2}{n_1^2 - n_1^2\sin^2\theta_1}}\right). \tag{1.66}$$

The above formula indicates that the reflection coefficient r_{TE} has an unit amplitude, so the intensity of the reflected and incident laser is equal, but a relative phase between these two light beams has been changed by $-2\Phi_{TE}$.

(2) TM polarization

According to Eq. (1.57), the reflection coefficients are given by

$$\begin{aligned} r_{\mathrm{TM}} &= \frac{n_2 \cos\theta_1 - n_1 \cos\theta_2}{n_2 \cos\theta_1 + n_1 \cos\theta_2} \\ &= \frac{\frac{n_2}{n_1}\sqrt{n_1^2 - n_1^2 \sin^2\theta_1} - \frac{n_1}{n_2}\sqrt{n_2^2 - n_1^2 \sin^2\theta_1}}{\frac{n_2}{n_1}\sqrt{n_1^2 - n_1^2 \sin^2\theta_1} + \frac{n_1}{n_2}\sqrt{n_2^2 - n_1^2 \sin^2\theta_1}} \\ &= \frac{\frac{n_2}{n_1}\sqrt{n_1^2 - n_1^2 \sin^2\theta_1} - i\frac{n_1}{n_2}\sqrt{n_1^2 \sin^2\theta_1 - n_2^2}}{\frac{n_2}{n_1}\sqrt{n_1^2 - n_1^2 \sin^2\theta_1} + i\frac{n_1}{n_2}\sqrt{n_1^2 \sin^2\theta_1 - n_2^2}} \\ &= \exp(-i2\Phi_{\mathrm{TM}}), \end{aligned} \tag{1.67}$$

where

$$\Phi_{\mathrm{TM}} = \tan^{-1}\left[\left(\frac{n_1}{n_2}\right)^2 \sqrt{\frac{n_1^2 \sin^2\theta_1 - n_2^2}{n_1^2 - n_1^2 \sin^2\theta_1}}\right]. \tag{1.68}$$

In Fig. 1.6, the half phase shift Φ_{TE} at a total internal reflection as a function of the incident angle θ_1 is plotted. As the incident angle increases from the critical angle to the grazing angle $\theta_1 = 90°$, the phase change Φ_{TE} also changes from zero to $\pi/2$. The derivative of the curve has an infinite derivative at the critical angle $\theta_1 = \theta_c$, but approaches $\left(1 - n_2^2/n_1^2\right)^{-1/2}$ when the incident angle $\theta_1 = 90°$. The curve of the phase change Φ_{TM} has a similar variation, which is also clear from the figure.

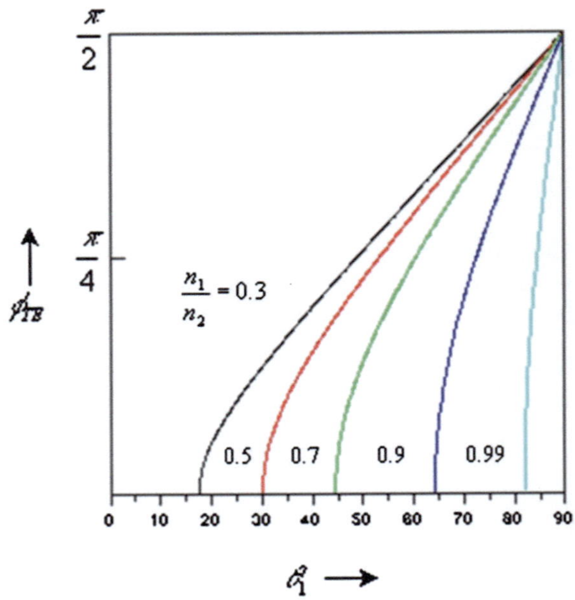

Fig. 1.6 Φ_{TE} as a function of θ_1

1.2.5 Goos–Hänchen Shift

In the above discussion, it was assumed that the light is reflected at the point of incidence, which only induces a phase shift. But in fact this assumption is not exact, and in 1947, Goos and Hänchen demonstrated experimentally that the light undergoes a small lateral shift at a total internal reflection, which can be evaluated by the following formula quantitatively:

$$D = \frac{cn_2\lambda}{(n_1^2 \sin^2\theta - n_2^2)^{1/2}}, \tag{1.69}$$

where c is a constant, if $n_1 = 1.52$ and $n_2 = 1$, then there is $c = 0.52$, and λ denotes the wavelength. The physics behind is that an actual laser beam has a certain spatial spectrum width, so that a finite sized beam is the superposition of infinite plane waves with a slightly different incident angle, and these plane waves interfere with each other.

In order to determine the amount of the lateral shift $2z_s$, we imagine a simple wave packet including just two plane waves with a different angle of incidence. If the z-component of the corresponding wave vector is $\beta \pm \Delta\beta$, respectively, the complex amplitude of the incident wave packet at the boundary $x = 0$ can be written as

$$\begin{aligned} A(z) &= [\exp(i\Delta\beta z) + \exp(-i\Delta\beta z)]\exp(i\beta z) \\ &= 2\cos(\Delta\beta z)\exp(i\beta z). \end{aligned} \tag{1.70}$$

Instead of applying the reflection law to each plane wave, we choose to expand the phase shift of reflection. Since the phase shift is a function of incident angle θ or β, when both $\Delta\varphi$ and $\Delta\beta$ are very small, it is simple to approximate the φ via the differential formula

$$\varphi(\beta \pm \Delta\beta) = \varphi(\beta) \pm \frac{\mathrm{d}\varphi}{\mathrm{d}\beta}\Delta\beta. \tag{1.71}$$

It follows that the complex amplitude of the reflected wave packet at $x = 0$ is

$$\begin{aligned} B(z) &= \{\exp[i(\Delta\beta z - 2\Delta\varphi)] + \exp[-i(\Delta\beta z - 2\Delta\varphi)]\}\exp[i(\beta z - 2\varphi)] \\ &= 2\cos[\Delta\beta(z - 2z_s)]\exp[i(\beta z - 2\varphi)], \end{aligned} \tag{1.72}$$

where

$$z_s = \frac{d\varphi}{d\beta}, \tag{1.73}$$

is the simple formula for the lateral shift of the reflected wave packet.

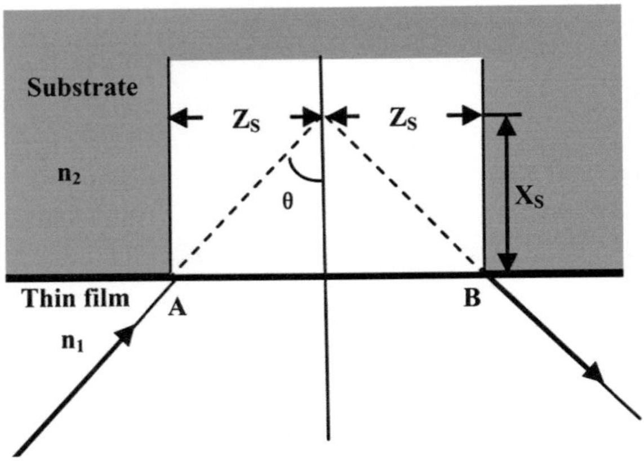

Fig. 1.7 Ray diagram of the total reflection at an interface

Using Eqs. (1.55) and (1.57), for the TE polarization, we have

$$k_0 z_s = \left(N^2 - n_2^2\right)^{-1/2} \tan\theta, \tag{1.74}$$

and for the TM mode, we have

$$k_0 z_s = \frac{\left(N^2 - n_2^2\right)^{-1/2} \tan\theta}{\left(\dfrac{N^2}{n_2^2} + \dfrac{N^2}{n_1^2} - 1\right)}. \tag{1.75}$$

As shown in Fig. 1.7, the light penetrates a finite amount into the second media before it returns to the first media, and the effective penetration depth x_s is given by

$$x_S = \frac{z_S}{\tan\theta}. \tag{1.76}$$

By comparing the result with the electromagnetic field solution which will be presented later, it will be clear that an evanescent field $e^{-\alpha_x x}$ exists in the substrate, and the effective penetration length is of the order of $1/\alpha_x$.

Chapter 2
Transfer Matrix Method and the Graded-Index Waveguide

Abstract The transfer matrix method used in thin-film optics is extremely useful when applied to analyze the propagation characteristics of electromagnetic waves in planar multilayer optical waveguides. This chapter aims to extend the transfer matrix method to treat the bound modes of the graded-index waveguide. Beginning with a brief introduction of the transfer matrix, we derived the eigenvalue equations and studied the multilayer optical waveguides. Different from the widely used WKB approximation, the transfer matrix obtained some important but different conclusions when applied to the graded-index waveguide, such as the exact phase shift at the classical turning points.

Keywords Transfer matrix method · Eigenvalue equation · WKB approximation · Graded-index waveguide

2.1 The Transfer Matrix and Its Characteristics

The 2×2 transfer matrix is a fruitful tool widely applied in optics to treat layered systems, such as superlattices or multilayered waveguide. And it is receiving more and more attention for its advantages such as easy computing and high accuracy. For example, it was used by M. Born and E. Wolf to investigate the transmission and reflection characteristics of light propagation through multilayer structures [1]. When dealing with multi-lens optical device or media, at each interface, the light is partially transmitted and partially reflected, and the matrix method can also provide good results [2]. In this section, we use some special solutions of the wave equation to construct a transfer matrix, which is a real matrix with clear physical insight. And in the rest of this chapter, the reader may find this method approachable and intriguing.

In optics, the regions with variable refractive index are usually approximated as a series of steps at a group of points, and between two adjacent points, the refractive index is treated as constant. So the polarization transfer matrix of the TE and TM

X. Wang et al., *Progress in Planar Optical Waveguides*,
Springer Tracts in Modern Physics 266, DOI 10.1007/978-3-662-48984-0_2

mode can be derived from the one-dimensional scalar wave equation to characterize the optical properties of these thin segments.

Considering the refractive index profile $n(x)$ of arbitrary shape as plotted in Fig. 2.1, without loss of generality, and we divide the region between the points of $x = a$ and $x = b$ into l subregions, and the width of each subregion is given by

$$w_j = x_j - x_{j-1} \quad (j = 1, 2, \ldots, l), \tag{2.1}$$

which becomes smaller with increasing l. In that case, the refractive index in the subregion can be viewed as homogeneous, and its strength is given by

$$n_j = n\left(\frac{x_{j-1} + x_j}{2}\right) \quad (j = 1, 2, \ldots, l). \tag{2.2}$$

Take the TE mode, for example, and let $\psi_j(x)$ denotes any field component of the electromagnetic distribution in the jth subregion (x_{j-1}, x_j), which satisfies the following scalar wave equation

$$\frac{d^2\psi_j(x)}{dx^2} + \kappa_j^2(x)\psi_j(x) = 0 \quad (j = 1, 2, \ldots, l), \tag{2.3}$$

with $\kappa_j^2(x) = k_0^2 n_j^2 - \beta^2$. Here, β is the propagation constant and κ_j denotes the wave number. At the interface $x = x_{j-1}$ between the $(j-1)$th and jth subregions, the continuity conditions of the wave function requires that

$$\begin{bmatrix} \psi_j(x_{j-1}) \\ \psi'_j(x_{j-1}) \end{bmatrix} = \begin{bmatrix} \psi_{j-1}(x_{j-1}) \\ \psi'_{j-1}(x_{j-1}) \end{bmatrix}. \tag{2.4}$$

Solving Eq. (2.3), the wave function in the jth subregion has the following form

$$\begin{cases} \psi_j(x) = A_j e^{i\kappa_j x} + B_j e^{-i\kappa_j x} \\ \psi'_j(x) = i\kappa_j (A_j e^{i\kappa_j x} - B_j e^{-i\kappa_j x}) \end{cases}, \tag{2.5}$$

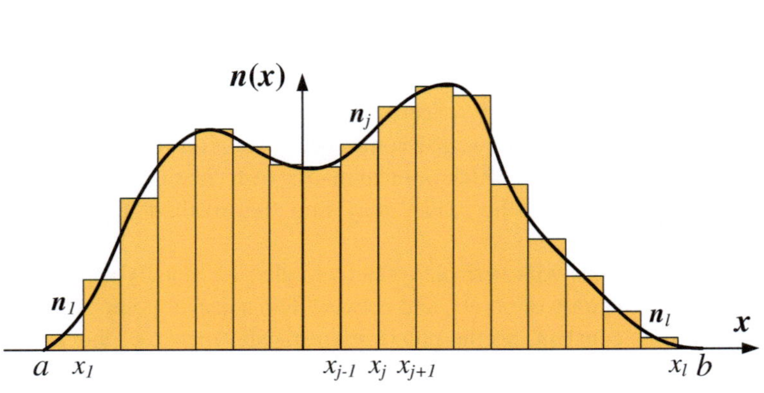

Fig. 2.1 One-dimensional refractive index profile of arbitrary distribution

which can be recast into a matrix form

$$\begin{bmatrix} \psi_j(x) \\ \psi_j'(x) \end{bmatrix} = \begin{bmatrix} e^{i\kappa_j x} & e^{-i\kappa_j x} \\ i\kappa_j e^{i\kappa_j x} & -i\kappa_j e^{-i\kappa_j x} \end{bmatrix} \begin{bmatrix} A_j \\ B_j \end{bmatrix}. \tag{2.6}$$

So at the point $x = x_j$, there is

$$\begin{bmatrix} \psi_j(x_j) \\ \psi_j'(x_j) \end{bmatrix} = \begin{bmatrix} e^{i\kappa_j x_j} & e^{-i\kappa_j x_j} \\ i\kappa_j e^{i\kappa_j x_j} & -i\kappa_j e^{-i\kappa_j x_j} \end{bmatrix} \begin{bmatrix} A_j \\ B_j \end{bmatrix}, \tag{2.7}$$

and at $x = x_{j-1}$, we can write down a similar matrix equation as follows:

$$\begin{bmatrix} \psi_j(x_{j-1}) \\ \psi_j'(x_{j-1}) \end{bmatrix} = \begin{bmatrix} e^{i\kappa_j x_{j-1}} & e^{-i\kappa_j x_{j-1}} \\ i\kappa_j e^{i\kappa_j x_{j-1}} & -i\kappa_j e^{-i\kappa_j x_{j-1}} \end{bmatrix} \begin{bmatrix} A_j \\ B_j \end{bmatrix}. \tag{2.8}$$

Combining Eqs. (2.7) and (2.8) yields

$$\begin{bmatrix} \psi_j(x_j) \\ \psi_j'(x_j) \end{bmatrix} = \begin{bmatrix} e^{i\kappa_j x_j} & e^{-i\kappa_j x_j} \\ i\kappa_j e^{i\kappa_j x_j} & -i\kappa_j e^{-i\kappa_j x_j} \end{bmatrix} \begin{bmatrix} e^{i\kappa_j x_{j-1}} & e^{-i\kappa_j x_{j-1}} \\ i\kappa_j e^{i\kappa_j x_{j-1}} & -i\kappa_j e^{-i\kappa_j x_{j-1}} \end{bmatrix}^{-1} \begin{bmatrix} \psi_j(x_{j-1}) \\ \psi_j'(x_{j-1}) \end{bmatrix}. \tag{2.9}$$

Through some basic matrix operations, Eq. (2.9) becomes

$$\begin{bmatrix} \psi_j(x_j) \\ \psi_j'(x_j) \end{bmatrix} = M_j \begin{bmatrix} \psi_j(x_{j-1}) \\ \psi_j'(x_{j-1}) \end{bmatrix}, \tag{2.10}$$

where

$$M_j = \begin{bmatrix} \cos(\kappa_j w_j) & \frac{1}{\kappa_j}\sin(\kappa_j w_j) \\ -\kappa_j \sin(\kappa_j w_j) & \cos(\kappa_j w_j) \end{bmatrix} \quad (j = 1, 2, \ldots, l). \tag{2.11}$$

and $w_j = x_j - x_{j-1}$ is the width of the subregion. Equation (2.11) is known as the transfer matrix in the subregion (x_{j-1}, x_j), which connects the wave function and its first derivative at the two boundaries of the jth subregion. And according to the boundary condition Eq. (2.4), the wave function and derivative at the boundary of the jth subregion are further connected with those at the $(j-1)$th subregion's boundary by

$$\begin{bmatrix} \psi_j(x_j) \\ \psi_j'(x_j) \end{bmatrix} = M_j \begin{bmatrix} \psi_{j-1}(x_{j-1}) \\ \psi_{j-1}'(x_{j-1}) \end{bmatrix}. \tag{2.12}$$

Take the TM mode into consideration, and the more generalized transfer matrix has the following form

$$M(w) = \begin{bmatrix} \cos(\kappa w) & \frac{f}{\kappa}\sin(\kappa w) \\ -\frac{\kappa}{f}\sin(\kappa w) & \cos(\kappa w) \end{bmatrix}, \tag{2.13}$$

where

$$f = \begin{cases} 1, & (TE) \\ n^2 & (TM) \end{cases}. \tag{2.14}$$

Before embarking on complicated issues, it is necessary to provide some discussion on the basic characteristics of the transfer matrix. For convenience, the 2×2 transfer matrix is rewritten as follows:

$$M = \begin{bmatrix} m_{11} & m_{12} \\ m_{21} & m_{22} \end{bmatrix}.$$

(a) Combining Eqs. (2.13) and (2.14), it is easy to found that in a non-absorptive medium, the matrix is a unimodular matrix with real coefficient

$$\det(M) = \begin{vmatrix} m_{11} & m_{12} \\ m_{21} & m_{22} \end{vmatrix} = m_{11}m_{22} - m_{12}m_{21} = 1, \tag{2.15}$$

where "det" represents a determinant. The physical insight of Eq. (2.15) is the conservation of energy.

(b) The energy eigenvalues λ of the transfer matrix can be determined via the secular equation

$$|M - \lambda I| = 0, \tag{2.16}$$

where I denotes the units matrix. Solving Eq. (2.16), and note that the modulus of the matrix equals unit, one can obtain the following equation:

$$\lambda^2 - (m_{11} + m_{22})\lambda + 1 = 0. \tag{2.17}$$

Equation (2.17) shows that the two eigenvalues of the matrix λ_1 and λ_2 are reciprocal to each other. Generally, the two eigenvalue can be expressed as follows:

$$\begin{cases} \lambda_1 = e^{i\kappa w} \\ \lambda_2 = e^{-i\kappa w} \end{cases}, \tag{2.18}$$

where the physics behind κ and h is determined by the specific structure. And according to Eq. (2.17), apparently there is

$$\cos(\kappa w) = \frac{1}{2}(m_{11} + m_{22}) = \frac{1}{2}\mathrm{Tr}M(w), \tag{2.19}$$

where "Tr" denotes the trace of the matrix. Equation (2.19) is an important formula for studying periodic structures, which is intimately connected with the Bloch theorem.

(c) Consider a stepped double potential well, and let $M(w_1)$ and $M(w_2)$ be the transfer matrix of the two adjacent homogeneous wells, respectively. According to Eq. (2.12), we have

$$\begin{bmatrix} \psi(w_1) \\ \psi'(w_1) \end{bmatrix} = M(w_1)\begin{bmatrix} \psi(0) \\ \psi'(0) \end{bmatrix} \tag{2.20}$$

and

$$\begin{bmatrix} \psi(w_1 + w_2) \\ \psi'(w_1 + w_2) \end{bmatrix} = M(w_2)\begin{bmatrix} \psi(w_1) \\ \psi'(w_1) \end{bmatrix}. \tag{2.21}$$

So there is

$$\begin{bmatrix} \psi(w_1 + w_2) \\ \psi'(w_1 + w_2) \end{bmatrix} = M(w_1 + w_2)\begin{bmatrix} \psi(0) \\ \psi'(0) \end{bmatrix}, \tag{2.22}$$

where

$$M(w_1 + w_2) = M(w_2)M(w_1). \tag{2.23}$$

Note that $M(w_2)$ and $M(w_1)$ in Eq. (2.23) cannot be swapped. The above consequent can be extended immediately to multilayer structure. Assume the respective width of a N-layer structure is $w_1, w_2, \ldots, w_N$, the corresponding matrixes of these homogeneous layers are $M(w_1), M(w_2), \ldots, M(w_N)$, and the transfer matrix of the whole structure is

$$M(w_1 + w_2 + \cdots + w_N) = M(w_N)M(w_{N-1})\cdots\cdots M(w_2)M(w_1). \tag{2.24}$$

(d) Periodic refractive index distribution, i.e., one-dimensional photonic crystal is common but extremely important. If the lattice length is Λ, and the transfer matrix for a single cell is

$$M(\Lambda) = \begin{bmatrix} m_{11} & m_{12} \\ m_{21} & m_{22} \end{bmatrix}, \tag{2.25}$$

then we can write down the transfer matrix for the whole lattice as

$$M(N\Lambda) = \underbrace{M(\Lambda) \cdot M(\Lambda) \cdots M(\Lambda)}_{N \text{ times}} = [M(\Lambda)]^N. \quad (2.26)$$

It is easy to prove from the above formula

$$[M(\Lambda)]^N = U_{N-1}(\chi)M(\Lambda) - U_{N-2}(\chi)E, \quad (2.27)$$

where $U_N(\chi)$ denotes the second-class Chebyshev polynomial

$$U_N(\chi) = \frac{\sin[(N+1)\arccos\chi]}{\sqrt{1-\chi^2}}. \quad (2.28)$$

Equation (2.26) can also be written as follows:

$$\begin{aligned} M(N\Lambda) &= [M(\Lambda)]^N \\ &= U_{N-1}(\chi)\begin{bmatrix} m_{11} & m_{12} \\ m_{21} & m_{22} \end{bmatrix} - U_{N-2}(\chi)\begin{bmatrix} 1 & 0 \\ 0 & 1 \end{bmatrix} \\ &= \begin{bmatrix} m_{11}U_{N-1}(\chi) - U_{N-2}(\chi) & m_{12}U_{N-1}(\chi) \\ m_{21}U_{N-1}(\chi) & m_{22}U_{N-1}(\chi) - U_{N-2}(\chi) \end{bmatrix}. \end{aligned} \quad (2.29)$$

By setting

$$M(\Lambda) = \begin{bmatrix} \cos\kappa\Lambda & \frac{f}{\kappa}\sin\kappa\Lambda \\ -\frac{\kappa}{f}\sin\kappa\Lambda & \cos\kappa\Lambda \end{bmatrix}, \quad (2.30)$$

we can recast Eq. (2.29) into

$$M(N\Lambda) = \begin{bmatrix} \cos(N\kappa\Lambda) & \frac{f}{\kappa}\sin(N\kappa\Lambda) \\ -\frac{\kappa}{f}\sin(N\kappa\Lambda) & \cos(N\kappa\Lambda) \end{bmatrix}. \quad (2.31)$$

(e) The inverse of the transfer matrix is defined by

$$MM^{-1} = I, \quad (2.32)$$

and for the transfer matrix, it is easy to find

$$M^{-1} = \begin{bmatrix} m_{22} & -m_{12} \\ -m_{21} & m_{11} \end{bmatrix}. \quad (2.33)$$

Using the inverse matrix, one can obtain the reverse transfer relationship. By multiplying $M^{-1}(h)$ on both sides of Eq. (2.12) yields

$$\begin{bmatrix} \psi_{j-1}(x_{j-1}) \\ \psi'_{j-1}(x_{j-1}) \end{bmatrix} = \begin{bmatrix} \cos(\kappa_j h_j) & -\frac{1}{\kappa_j}\sin(\kappa_j h_j) \\ \kappa_j \sin(\kappa_j h_j) & \cos(\kappa_j h_j) \end{bmatrix} \begin{bmatrix} \psi_j(x_j) \\ \psi'_j(x_j) \end{bmatrix}. \quad (2.34)$$

Since both the transfer matrix and its inverse can relate the wave function at two points, they are both referred as transfer matrix in the rest of the book, while the only difference is the different transfer direction.

(f) If we have $k_0^2 n^2 < \beta^2$ in a thin layer, then the solution of the scalar wave equation in this region is the superposition of two exponential functions, while the transverse wave number κ corresponding to oscillating field is replaced by an attenuation coefficient α, and there is

$$\kappa = i\alpha. \quad (2.35)$$

Note that

$$\begin{cases} \sin(ix) = i\sinh(x) \\ \cos(ix) = \cosh(x) \end{cases}, \quad (2.36)$$

Equation (2.13) should also be replaced by the following expression

$$M_j = \begin{bmatrix} \cosh(\alpha_j h_j) & \frac{1}{\alpha_j}\sinh(\alpha_j h_j) \\ \alpha_j \sinh(\alpha_j h_j) & \cosh(\alpha_j h_j) \end{bmatrix}, \quad (2.37)$$

while its inverse is

$$M_j^{-1} = \begin{bmatrix} \cosh(\alpha_j h_j) & -\frac{1}{\alpha_j}\sinh(\alpha_j \mathrm{h_j}) \\ -\alpha_j \sinh(\alpha_j h_j) & \cosh(\alpha_j \mathrm{h_j}) \end{bmatrix}. \quad (2.38)$$

2.2 The Eigenvalue Equation

Consider a simple planar waveguide, whose refractive index distribution is plotted in Fig. 2.2, and this section is aimed to calculate its eigenvalue spectrum. Since the transfer matrix, which connects the wave function and its first derivative at the two interfaces of a thin layer, represents the characteristic parameters of the dielectric slab, the field distribution in the guiding layer is not need to be considered. As a result, this procedure will be much simplified if the transfer matrix is applied, we only need to determine the wave function in the regions of $x < 0$ and $x > w$.

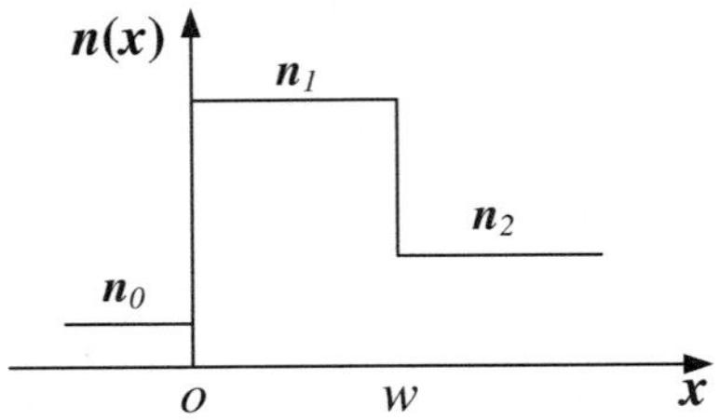

Fig. 2.2 Refractive index distribution of a planar waveguide with three dielectric layers

For the refractive index as plotted in Fig. 2.2, the transverse component of the TE mode transmitted in the guiding layer between the dielectric layer of n_0 and n_2 is as follows:

$$E_y(x) = \begin{cases} A\exp(p_0 x) & -\infty < x < 0 \\ D\exp[-p_2(x-w)] & w < x < +\infty \end{cases}, \tag{2.39}$$

and it follows

$$\begin{cases} E_y(0) = A \\ E_y'(0) = p_0 A \end{cases} \quad . \quad \begin{cases} E_y(h) = D \\ E_y'(h) = -p_2 D \end{cases} \tag{2.40}$$

Substituting the above equation into Eq. (2.12), one obtains

$$A\begin{bmatrix} 1 \\ p_0 \end{bmatrix} = \begin{bmatrix} \cos(\kappa_1 h) & -\frac{1}{\kappa_1}\sin(\kappa_1 \mathrm{h}) \\ \kappa_1 \sin(\kappa_1 h) & \cos(\kappa_1 \mathrm{h}) \end{bmatrix} \begin{bmatrix} 1 \\ -p_2 \end{bmatrix} D. \tag{2.41}$$

Multiply both sides of the above equation by a row vector $[-p_0\ 1]$, there is

$$[-p_0 \quad 1]\begin{bmatrix} \cos(\kappa_1 h) & -\frac{1}{\kappa_1}\sin(\kappa_1 h) \\ \kappa_1 \sin(\kappa_1 h) & \cos(\kappa_1 h) \end{bmatrix} \begin{bmatrix} 1 \\ -p_2 \end{bmatrix} = 0, \tag{2.42}$$

and by solving the above equation, it yields

$$\tan(\kappa_1 h) = \frac{p_0 + p_2}{\kappa_1\left(1 - \frac{p_0 p_2}{\kappa_1^2}\right)}. \tag{2.43}$$

Eq. (2.43) can be called as the mode eigenvalue equation of TE polarization. If we take the boundary condition of the TM mode into consideration, the mode eigenvalue equation for TM mode can be written as follows:

$$\begin{bmatrix} -\frac{p_0}{n_0^2} & 1 \end{bmatrix} \begin{bmatrix} \cos(\kappa_1 h) & -\frac{n_1^2}{\kappa_1}\sin(\kappa_1 h) \\ \frac{\kappa_1}{n_1^2}\sin(\kappa_1 h) & \cos(\kappa_1 h) \end{bmatrix} \begin{bmatrix} 1 \\ -\frac{p_2}{n_2^2} \end{bmatrix} = 0, \tag{2.44}$$

and the equivalent phase-type dispersion equation by solving Eq. (2.44) is as follows:

$$\tan(\kappa_1 h) = \frac{n_1^2 \kappa_1 (n_2^2 p_0 + n_0^2 p_2)}{n_0^2 n_2^2 \kappa_1^2 - n_1^4 p_0 p_2}. \tag{2.45}$$

2.3 WKB Approximation [3, 4]

Before we embark on the bound states of the graded-index optical waveguide, we first review briefly the widely applied WKB approximation, which has also found wide application in optical waveguide theory. Let us begin with the mathematical derivation of the WKB wave function, and consider again the solutions of the scalar wave equation:

$$\frac{\mathrm{d}^2\psi(x)}{\mathrm{d}x^2} + \left(k^2 n^2(x) - \beta^2\right)\psi(x) = 0. \tag{2.46}$$

If we suppose the potential varies very slowly, then we can write the trial solution as a combination of two plane waves traveling leftward and rightward, respectively:

$$\psi(x) = A(0)\exp(ikS(x)) + B(0)\exp(-ikS(x)), \tag{2.47}$$

which is a good approximation only when the refractive index varies very slowly. How slow? For simplicity, inserting a rightward traveling plane wave into the wave equation, one will get a differential equation of $S(x)$:

$$\left(\frac{\mathrm{d}S}{\mathrm{d}x}\right)^2 + \frac{1}{ik}\frac{\mathrm{d}^2 S}{\mathrm{d}x^2} = n^2(x) - \frac{\beta^2}{k^2}. \tag{2.48}$$

We now expand $S(x)$ in the power series in $1/k$ and write as follows:

$$S = S_0 + \frac{1}{k}S_1 + \left(\frac{1}{k}\right)^2 S_2 + \cdots. \tag{2.49}$$

Feeding this into Eq. (2.48) and requiring that all terms of $O(\hbar^n)$ vanish independently, there is

$$\left|\frac{\mathrm{d}\lambda}{\mathrm{d}x}\right| \ll 1. \tag{2.50}$$

where

$$\lambda = \frac{2\pi}{\sqrt{k^2 n(x)^2 - \beta^2}}. \tag{2.51}$$

Equation (2.50) requires that the refractive index should vary slowly, and there is

$$\left|\frac{n(x)\lambda}{2\pi\left(n(x)^2 - \beta^2/k^2\right)} \frac{\mathrm{d}n(x)}{\mathrm{d}x}\right| \ll 1. \tag{2.52}$$

If Eq. (2.52) is satisfied, and we ignore all terms of $O(1/k^n)(n \geq 2)$ in Eq. (3.4), we can rewrite down S_0, S_1 as follows:

$$\begin{cases} S_0 = \frac{1}{k}\int (k^2 n(x)^2 - \beta^2)^{1/2}\mathrm{d}x \\ S_1 = \frac{i}{2}\ln\left|\frac{\mathrm{d}s_0}{\mathrm{d}x}\right| \end{cases}, \tag{2.53}$$

and the first-order WKB wave function Eq. (2.47) as follows:

$$\begin{cases} \psi(x) = \frac{A}{\sqrt{\kappa}}\exp\left[i\int^x \kappa \mathrm{d}x\right] + \frac{B}{\sqrt{\kappa}}\exp\left[-i\int^x \kappa \mathrm{d}x\right] & \kappa^2 = k^2 n^2(x) - \beta^2 > 0 \\ \psi(x) = \frac{A}{\sqrt{p}}\exp\left[i\int^x p\mathrm{d}x\right] + \frac{B}{\sqrt{p}}\exp\left[-i\int^x p\mathrm{d}x\right] & p^2 = \beta^2 - k^2 n^2(x) > 0 \end{cases}, \tag{2.54}$$

Equation (2.54) shows that we should anticipate an oscillatory behavior in a region where $k^2n^2(x) > \beta^2$ and an evanescent behavior in the opposite region. In the region of turning point given by $kn(x) = \beta$, the WKB wave function breaks down since Eq. (2.52) is no longer fulfilled. In order to construct a globally WKB wave function, connection formulas at turning points are required to match to the WKB solutions on both sides of the turning point regions, where the local wavelength λ is singular. Here, we present the connection formulas directly, and the mathematics progresses are referred to the related references. Suppose the position of the turning point is given by x_t, and let us identify the region $k^2n^2(x) > \beta^2$ with $x < x_t$ and vice versa. The corresponding connecting formula is as follows:

$$\frac{1}{2\sqrt{p}}\exp\left(-\frac{1}{\hbar}\int_{x_t}^{x} p(x)dx\right) \leftrightarrow \frac{1}{\sqrt{\kappa}}\cos\left(\frac{1}{\hbar}\int_{x}^{x_t} p(x)\mathrm{d}x - \frac{\pi}{4}\right) \tag{2.55}$$
$$\left(k^2n^2(x) < \beta^2\right) \qquad\qquad \left(k^2n^2(x) > \beta^2\right).$$

However, in some issues that the wave function in the optically dense media is not the superposition of two waves propagate in the opposite directions, such as the transmitted waves left the optically sparse media and traveled to infinite, the connection formula should be replaced by

$$\begin{aligned} &\frac{1}{\sqrt{p}}\exp\left(\int_{x}^{x_t} p(x)\mathrm{d}x\right) \leftrightarrow -\frac{1}{\sqrt{\kappa}}\exp\left(i\int_{x_t}^{x} \kappa \mathrm{d}x + \frac{i\pi}{4}\right) \\ &\left(k^2n^2(x) < \beta^2\right) \qquad \left(k^2n^2(x) > \beta^2\right). \end{aligned} \tag{2.56}$$

Instead of the two conventional expressions above, Prof. Friedrich [5] proposed that the application of the WKB approximation can be significantly extended if the connection formulas in the most general case can be used. These expressions can be written as follows:

$$\frac{2}{\sqrt{p(x)}}\cos\left(\int_{x_t}^{x} p(x)\mathrm{d}x - \frac{\phi}{2}\right) \leftrightarrow \frac{N}{\sqrt{p}}\exp\left(-\int_{x}^{x_t} p\mathrm{d}x\right), \tag{2.57}$$

$$\frac{1}{\sqrt{\kappa}}\cos\left(\int_{x}^{x_t} \kappa \mathrm{d}x - \frac{\bar{\phi}}{2}\right) \leftrightarrow \frac{\bar{N}}{\sqrt{p}}\exp\left(\int_{x_t}^{x} p(x)\mathrm{d}x\right). \tag{2.58}$$

There are four parameters $N, \overline{N}, \phi$, and $\bar{\phi}$ to be determined by considering the specific problems. And the conventional formulas can be retrieved by setting $N = 1$ and $\phi = \pi/2$. If we consider a superposition of the above two expressions $\psi = A \times (3.39) + B \times (3.40)$ with arbitrary complex coefficients A and B, the conservation condition of the current density on the two sides of the turning point requires that

$$N\bar{N} = \sin\left(\frac{\phi - \bar{\phi}}{2}\right), \tag{2.59}$$

which can be used to determine the undetermined parameters. Imagine a refractive index profile that varies slowly, where the two turning points x_{tl} and x_{tr} are defined via $n(x_{tl}) = n(x_{tr}) = \beta/\kappa$, so there will be oscillatory behavior in the region $x_{tl} < x < x_{tr}$ and evanescent behavior elsewhere. The WKB waves in the region $x_{tl} < x < x_{tr}$ can be defined as follows:

$$\psi_{WKB}(x) \propto \frac{1}{\sqrt{\kappa(x)}}\cos\left(\int_{x_{tl}}^{x} \kappa(x')\mathrm{d}x' - \frac{\phi_l}{2}\right), \tag{2.60}$$

from the left turning point, or equivalently

$$\psi_{WKB}(x) \propto \frac{1}{\sqrt{\kappa(x)}}\cos\left(\int_{x}^{x_{tr}} \kappa(x')\mathrm{d}x' - \frac{\phi_r}{2}\right), \tag{2.61}$$

from the right turning point. ϕ_l and ϕ_r are the reflection phases at the left and right turning points, respectively, and x is an arbitrary point in the well away from the turning points. According to connection expression (2.55), we have $\phi_l = \phi_r = \pi/2$ here. So the two expressions above must agree with each other, and this requirement can only be satisfied when the sum of the two arguments equals an integral multiple of π, which yields the following:

$$\int_{x_{tl}}^{x_{tr}} \kappa(x)\mathrm{d}x = (n + \frac{1}{2})\pi, \tag{2.62}$$

where $n = 0, 1, 2, \ldots$ The above expression is the famous WKB resonance condition and may be used to find the eigenvalue equation for a graded-index waveguide with n as the mode number. The above formula is valid, provided that the two turning points are positioned sufficiently far apart.

In conclusion, in this section, we review briefly the semiclassical WKB approximation, including its wave function and the eigenvalue equation (quantization condition). And it should be note that the basic WKB wave function ignores all terms of $O(1/k^n) \quad (n \geq 2)$, and the phase shift ϕ_l, ϕ_r at the turning points in the original WKB approximation is equal to $\pi/2$.

2.4 Multilayer Optical Waveguides

2.4.1 Asymmetric Four-Layer Slab Waveguide [6]

In the beginning of this section, we use a simple asymmetric slab waveguide to demonstrate the existence of the scattered subwaves, which can be simply defined as follows: All the waves being reflected at non-classical turning points for at least once are referred as the scattered subwaves. In contrast, the waves that only reflected at classical turning points are called as the main waves. This is the first time we proposed the concept of the scattered subwaves in this book. Like many other basic concepts, the seemingly simple concept of scattered subwaves is not as straightforward as one might assume.

We are now in a position to deal with the four-layer dielectric slab waveguide via the transfer matrix method. As plotted in Fig. 2.3, two uniform isotropic dielectric of refractive indexes n_1 and n_2 and thicknesses h_1 and h_2, are sandwiched between two semi-infinite layers of lower index n_0 and n_3. For definiteness,

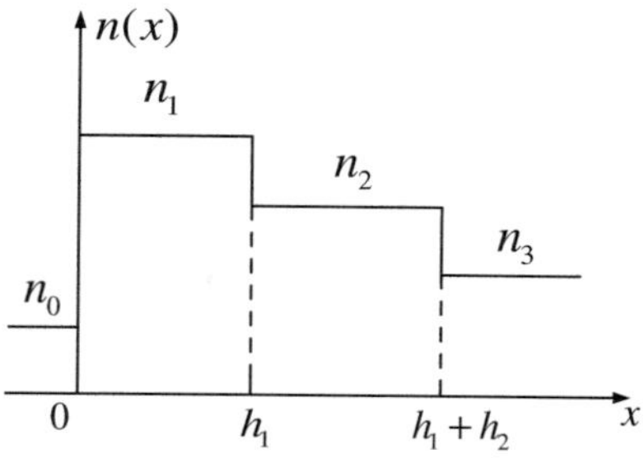

Fig. 2.3 The refractive index distribution of a four-layer dielectric slab waveguide

considering the situation $n_1 > n_2 > n_3 > n_0$, which yields an asymmetric guiding structure, and we are interested in those guide modes, whose power is confined largely to the central layer of the guide. So in this chapter, we only consider two cases for the propagation constant β, including (A) $kn_2 \geq \beta \geq kn_3$, that both regions 1 and 2 are the regions of electromagnetic confinement; (B) $kn_1 \geq \beta \geq kn_2$, that only region 1 is the primary region of energy confinement. The discussion on the case of leaky waves is left to the fourth chapter.

For case A of $kn_2 \geq \beta \geq kn_3$, the eigenvalue equation of the matrix form can be immediately written as follows:

$$\begin{bmatrix} -p_0 & 1 \end{bmatrix} M_1 M_2 \begin{bmatrix} 1 \\ -p_3 \end{bmatrix} = 0, \tag{2.63}$$

where

$$M_1 = \begin{bmatrix} \cos(\kappa_1 h_1) & -\frac{1}{k_1}\sin(\kappa_1 h_1) \\ \kappa_1 \sin(\kappa_1 h_1) & \cos(\kappa_1 h_1) \end{bmatrix}, \tag{2.64}$$

$$M_2 = \begin{bmatrix} \cos(\kappa_2 h_2) & -\frac{1}{k_2}\sin(\kappa_2 h_2) \\ \kappa_2 \sin(\kappa_2 h_2) & \cos(\kappa_2 h_2) \end{bmatrix}, \tag{2.65}$$

$$\left.\begin{aligned} \kappa_1 &= (k_0^2 n_1^2 - \beta^2)^{1/2} \\ \kappa_2 &= (k_0^2 n_2^2 - \beta^2)^{1/2} \\ p_0 &= (\beta^2 - k_0^2 n_0^2)^{1/2} \\ p_3 &= (\beta^2 - k_0^2 n_3^2)^{1/2} \end{aligned}\right\}. \tag{2.66}$$

Substituting the matrixes into Eq. (2.63), the eigenvalue equation of the asymmetric four-layer slab waveguide is as follows:

$$\kappa_1 h_1 = m\pi + \tan^{-1}\left(\frac{p_0}{\kappa_1}\right) + \tan^{-1}\left(\frac{p_2}{\kappa_1}\right), \quad (m = 0, 1, 2, \ldots), \tag{2.67}$$

where

$$p_2 = \kappa_2 \tan\left[\tan^{-1}\left(\frac{p_3}{\kappa_2}\right) - \kappa_2 h_2\right]. \tag{2.68}$$

In order to see the physical insight of the above expression, we define a new quantity Φ_2, which is given by the following:

$$\Phi_2 = \tan^{-1}\left(\frac{p_2}{\kappa_2}\right), \tag{2.69}$$

and which can be rewritten in the following form according to Eq. (2.68):

$$\kappa_2 h_2 + \Phi_2 = m'\pi + \tan^{-1}\left(\frac{p_3}{\kappa_2}\right), \quad (m' = 0, 1, 2, \ldots). \tag{2.70}$$

Combine Eqs. (2.69) and (2.70) with the equation below:

$$\tan^{-1}\left(\frac{p_2}{\kappa_1}\right) = \tan^{-1}\left[\frac{\kappa_2}{\kappa_1}\tan(\Phi_2)\right]. \tag{2.71}$$

We finally obtain an eigenvalue equation that has a similar form of the three-layer slab waveguide

$$\kappa_1 h_1 + \kappa_2 h_2 + \Phi(s) = m\pi + \tan^{-1}\left(\frac{p_0}{\kappa_1}\right) + \tan^{-1}\left(\frac{p_3}{\kappa_2}\right), \quad (m = 0, 1, 2, \ldots) \tag{2.72}$$

where

$$\Phi(s) = \Phi_2 - \tan^{-1}\left(\frac{\kappa_2}{\kappa_1}\tan\Phi_2\right). \tag{2.73}$$

To clarify the physics behind the unknown $\Phi(s)$, let us consider the special case of $\frac{n_1^2 - n_2^2}{n_1^2} \ll 1$, which follows that $\frac{\kappa_1 - \kappa_2}{\kappa_1} \ll 1$. Using differential formula, there is

$$\begin{aligned} \tan^{-1}\left(\frac{\kappa_2}{\kappa_1}\tan\Phi_2\right) &= \tan^{-1}\left[\left(1 - \frac{\kappa_1 - \kappa_2}{\kappa_1}\right)\tan\Phi_2\right] \\ &\approx \Phi_2 - \frac{\kappa_1 - \kappa_2}{2\kappa_1}\sin 2\Phi_2, \end{aligned} \tag{2.74}$$

and according to Eq. (2.73), one can obtain

$$\Phi(s) = \frac{\kappa_1 - \kappa_2}{2\kappa_1} \sin 2\Phi_2. \tag{2.75}$$

The amplitude of the right-hand side of Eq. (2.75) under first-order approximation is as follows:

$$\frac{\kappa_1 - \kappa_2}{2\kappa_1} \approx \frac{\kappa_1 - \kappa_2}{\kappa_1 + \kappa_2}, \tag{2.76}$$

which denotes the reflection coefficient of light incident from region 1 to region 2. Consequently, $\Phi(s)$ can be viewed as the reflection phase contribution of the first-order scattered subwaves. Of course, the term $\Phi(s)$ denotes the phase contribution of all the scattered subwaves if we did not carry out any approximation. On the contrary, if $n_1 = n_2$ holds, there is $\Phi(s) = 0$. As a summary, $\Phi(s)$ is the phase contribution induced by the reflection occurs at the interface between regions 1 and 2, and is determined by the difference of the refractive index between the two regions. So when dealing with multilayer waveguide, both phase contribution of the main waves and the scattered subwaves should be taken into consideration.

For case B $k_0 n_1 > \beta > k_0 n_2$, whose guiding layer locates in the region of $(0, h_1)$. In this case, the matrix form of the eigenvalue equation can still be written as Eq. (2.63), but the κ_2 in Eq. (2.65) should be modified as follows:

$$\kappa_2 = i\left(\beta^2 - k_0^2 n_2^2\right)^{1/2} = i\alpha_2. \tag{2.77}$$

Consequently, the sine and cos in the matrix should be replaced as sinh and cosh, that is,

$$\sin(\kappa_2 h_2) = i \sinh(\alpha_2 h_2), \quad \cos(\kappa_2 h_2) = \cosh(\alpha_2 h_2). \tag{2.78}$$

And the transfer matrix M_2 becomes

$$M_2 = \begin{bmatrix} \cosh(\alpha_2 h_2) & -\frac{1}{\alpha_2}\sinh(\alpha_2 h_2) \\ -\alpha_2 \sinh(\alpha_2 h_2) & \cosh(\alpha_2 h_2) \end{bmatrix}. \tag{2.79}$$

So the eigenvalue equation for this case is mathematically the same with Eq. (2.67), except that p_2 is defined by

$$p_2 = \alpha_2 \tanh\left[\tanh^{-1}\left(\frac{p_3}{\alpha_2}\right) + \alpha_2 h_2\right]. \tag{2.80}$$

To see the scattered subwaves in the four-layer slab waveguide in another way, let us reconsider the case A for TE mode, whose transverse electric field may be expressed as follows:

$$E_y(x) = \begin{cases} A_0 \exp(p_0 x) & -\infty < x < 0 \\ A_1 \exp(i\kappa_1 x) + B_1 \exp(-i\kappa_1 x) & 0 < x < h_1 \\ A_2 \exp[i\kappa_2(x - h_1)] + B_2 \exp[-i\kappa_2(x - h_1)] & h_1 < x < h_1 + h_2 \\ A_3 \exp[-p_3(x - h_1 - h_2)] & h_1 + h_2 < x < +\infty \end{cases} \tag{2.81}$$

where

$$\left.\begin{aligned} \kappa_1 &= (k_0^2 n_1^2 - \beta^2)^{1/2} \\ \kappa_2 &= (k_0^2 n_2^2 - \beta^2)^{1/2} \\ p_0 &= (\beta^2 - k_0^2 n_0^2)^{1/2} \\ p_3 &= (\beta^2 - k_0^2 n_3^2)^{1/2} \end{aligned}\right\}.$$

Using the continuity condition of E_y and $\partial E_y/\partial x$ at boundaries $x = 0$, $x = h_1$, and $x = h_1 + h_2$, one may write down the dispersion equation as follows:

$$\begin{aligned} &\exp[i2(\kappa_1 h_1 + \kappa_2 h_2 - \Phi_{10} - \Phi_{23})] + \frac{\kappa_1 - \kappa_2}{\kappa_1 + \kappa_2} \exp[i2(\kappa_1 h_1 - \Phi_{10})] \\ &\quad + \frac{\kappa_2 - \kappa_1}{\kappa_2 + \kappa_1} \exp[i2(\kappa_2 h_2 - \Phi_{23})] = 1, \end{aligned} \tag{2.82}$$

where

$$\Phi_{10} = \tan^{-1}\left(\frac{p_0}{\kappa_1}\right) \tag{2.83}$$

$$\Phi_{23} = \tan^{-1}\left(\frac{p_3}{\kappa_2}\right) \tag{2.84}$$

What does Eq. (2.82) means? See Fig. 2.4 for the zigzag path of rays in the slab waveguide. Clearly, the first term on the left-hand side of Eq. (2.82) denotes the main waves which are plotted with solid lines in Fig. 2.4. Starting from the interface between regions 0 and 1, the main waves travels through the interface between regions 1 and 2 and then is total-reflected at the boundary of region 3. In summary, the main waves only is total-reflected at the boundaries of regions 0 and 3. On the contrary, the dotted lines shown in Fig. 2.4 represent the second and third terms in Eq. (2.82), which are reflected at the interface between regions 1 and 2. So these two terms denote the scattered subwaves. The guided modes in the slab waveguide are in fact the coherent superposition result of the main waves and the scattered subwaves, which are ignored in the semiclassical theories. It can be proved mathematically that Eq. (2.82) can be recast as follows:

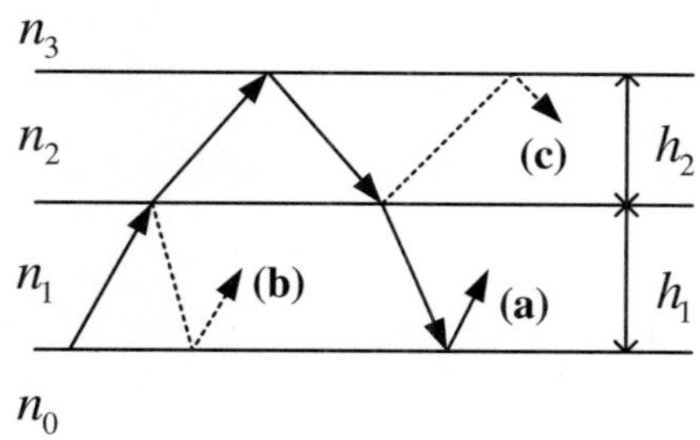

Fig. 2.4 Zigzag path of the main waves (*a*) and the scattered subwaves (*b*, *c*) in the four-layer dielectric waveguide

$$\exp\{i2[\kappa_1 h_1 + \kappa_2 h_2 + \Phi(s) - \Phi_{10} - \Phi_{23}]\} = 1. \tag{2.85}$$

In view of $\exp(i2m\pi) = 1,\ (m = 0, 1, 2, \ldots)$, the above equation is equivalent with Eq. (2.72).

2.4.2 Multilayer Slab Waveguide

In this section, we expand the conclusions of the four-layer slab waveguide to the multilayer slab waveguide. Suppose there are l layers of index n_1, n_2, … n_l, and thickness of h_1, h_2, … h_l, embedded in two cladding layers of index n_0 and n_{l+1}. For the structure consider here, there is $n_1 > n_2 > \ldots > n_l > n_{l+1} > n_0$ (Fig. 2.5).

Let us extend Eq. (2.63) to this waveguide structure, consider the guided modes with $kn_l > \beta > kn_{l+1}$, and the eigenvalue equation for the TE mode in the matrix form can be written as follows:

$$[-p_0 \quad 1]\prod_{i=1}^{l} M_i \begin{bmatrix} 1 \\ -p_{l+1} \end{bmatrix} = 0, \tag{2.86}$$

Fig. 2.5 Asymmetric $l + 2$ layer slab waveguide

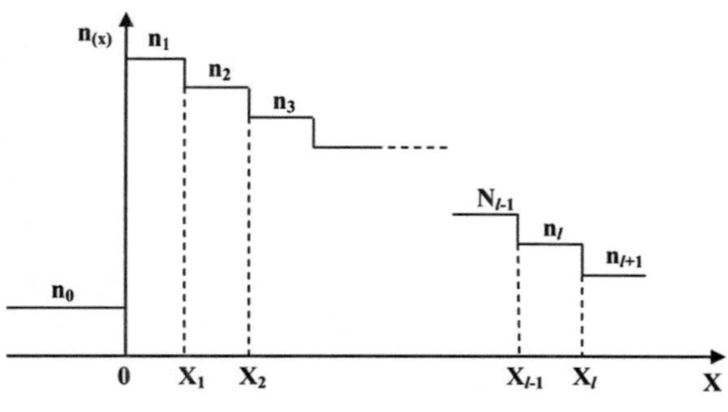

where the matrix corresponding to the ith layer $\mathrm{M_i}$ has the following form:

$$M_i = \begin{bmatrix} \cos(\kappa_i h_i) & -\frac{1}{\kappa_i}\sin(\kappa_i h_i) \\ \kappa_i \sin(\kappa_i h_i) & \cos(\kappa_i h_i) \end{bmatrix} \tag{2.87}$$

where

$$\begin{cases} \kappa_i = (k_0^2 n_i^2 - \beta^2)^{1/2} \\ p_0 = (\beta^2 - k_0^2 n_0^2)^{1/2} \\ p_{l+1} = (\beta^2 - k_0^2 n_{l+1}^2)^{1/2} \end{cases}.$$

Equation (2.86) can be simplified via direct algebraic manipulations as

$$\begin{gathered} \kappa_1 h_1 = m\pi + \tan^{-1}\left(\frac{p_0}{\kappa_1}\right) + \tan^{-1}\left(\frac{p_2}{\kappa_1}\right), \\ (m = 0, 1, 2, \ldots) \end{gathered} \tag{2.88}$$

where

$$\begin{gathered} p_i = \kappa_i \tan\left[\tan^{-1}\left(\frac{p_{i+1}}{\kappa_i}\right) - \kappa_i h_i\right] \\ i = (2, 3, \ldots, l) \end{gathered} \tag{2.89}$$

The two formula above completly specify the dispersion characteristics of the asymmetric multilayer slab waveguide. However, Eq. (2.89) is a recurrence formula, which requires all the information of $p_j (j > i)$ to calculate p_i. To see the different roles of the main waves and the scattered subwaves, we define

$$\phi_i = \tan^{-1}\left(\frac{p_i}{\kappa_i}\right), \tag{2.90}$$

According to Eq. (2.89), one obtains

$$\begin{aligned} \phi_i &= m_i\pi + \tan^{-1}\left(\frac{p_{i+1}}{\kappa_i}\right) - \kappa_i h_i \\ &= m_i\pi + \tan^{-1}\left(\frac{\kappa_{i+1}}{\kappa_i}\tan\phi_{i+1}\right) - \kappa_i h_i \\ &(m_i = 0, 1, 2 \ldots; i = 1, 2, \ldots, l-1), \end{aligned} \tag{2.91}$$

which can be modified as

$$\begin{gathered} \kappa_i h_i + \left[\phi_{i+1} - \tan^{-1}\left(\frac{\kappa_{i+1}}{\kappa_i}\tan\phi_{i+1}\right)\right] = m_i\pi + (\phi_{i+1} - \phi_i). \\ (m_i = 0, 1, 2, \ldots; i = 1, 2, \ldots, l-1) \end{gathered} \tag{2.92}$$

When $i = l$, there is

$$\kappa_l h_l = m_l \pi + \tan^{-1}\left(\frac{p_{l+1}}{\kappa_l}\right) - \varphi_l, \tag{2.93}$$

Based on the three equations above and sum up over i, one can write down

$$\begin{aligned} &\sum_{i=1}^{l} \kappa_i h_i + \sum_{i=1}^{l-1}\left[\phi_{i+1} - \tan^{-1}\left(\frac{\kappa_{i+1}}{\kappa_i}\tan\phi_{i+1}\right)\right] \\ &= m\pi + \tan^{-1}\left(\frac{p_{l+1}}{\kappa_l}\right) - \phi_l. \end{aligned} \tag{2.94}$$

On the other hand, it is easy to prove

$$\phi_1 = m_1\pi + \tan^{-1}\left(\frac{p_2}{\kappa_1}\right) - \kappa_1 h_1, \tag{2.95}$$

which can be rewritten by inserting Eq. (2.88)

$$\phi_1 = m_1\pi - \tan^{-1}\left(\frac{p_0}{\kappa_1}\right). \tag{2.96}$$

Finally, we transformed Eq. (2.94) into

$$\begin{gathered} \sum_{i=1}^{l} \kappa_i h_i + \Phi(s) = m\pi + \tan^{-1}\left(\frac{p_0}{\kappa_1}\right) + \tan^{-1}\left(\frac{p_{l+1}}{\kappa_l}\right), \\ (m = 0, 1, 2, \ldots) \end{gathered} \tag{2.97}$$

with the phase contribution of the scattered subwaves

$$\Phi(s) = \sum_{i=1}^{l-1}\left[\phi_{i+1} - \tan^{-1}\left(\frac{\kappa_{i+1}}{\kappa_i}\tan\phi_{i+1}\right)\right]. \tag{2.98}$$

We can obtain the eigenvalue equation of the three-layer or four-layer slab structure from Eq. (2.97) by setting $l = 1$, or $l = 2$, respectively. It is obvious that this formula can be applied to any multilayer structures without any approximation. Furthermore, it has a clear physical explanation that both the main waves and scattered subwaves contribute to the total phase contribution. Although the discussion above consider only the case of $\beta < k_0 n_l$, it is not difficult to obtain the corresponding result related to the case of $\beta > k_0 n_i$, by just replacing the κ_j with $i\alpha_j$ in the matrix M_j for all the $j > i$. Interesting readers can also derive the eigenvalue equation for TM modes by using the appropriate matrix.

2.5 The Transfer Matrix Treatment of the Graded-Index Waveguide

2.5.1 The Eigenvalue Equation

In this section, the transfer matrix method is extended to treat the graded-index waveguide, and it is demonstrated strictly that the phase shifts at the turning points are exact equal to π. Since this section deals with the general graded-index structure with arbitrary refractive index profile, the strategy is as follows: We first approximate the graded-index waveguide with a multilayer waveguide with n layers and then take the limit as the n approaches infinite. At the beginning, let us consider a simple case in which only one turning point exists (Fig. 2.6).

Assume that the turning point locates at the position $x = x_t$, and the index profile extends to infinite. In order to apply the transfer matrix method, we should truncate infinite at $x_s = x_t + x_c$ for sufficient large x_c, and set $n(x) = n_s$ for $x > x_s$. Then, the regions $(0, x_t)$ and (x_t, x_s) are divided into l and m segments, with each layer has the same thickness h, so that $x_t = lh$ and $x_c = mh$. For TE mode, the transfer matrix for these segments is as follows:

$$M_i = \begin{bmatrix} \cos(\kappa_i h) & -\frac{1}{\kappa_i}\sin(\kappa_i h) \\ \kappa_i \sin(\kappa_i h) & \cos(\kappa_i h) \end{bmatrix} \quad (i = 1, 2, \ldots l), \tag{2.99}$$

and

$$M_j = \begin{bmatrix} \cosh(\alpha_j h) & -\frac{1}{\alpha_j}\sinh(\alpha_j h) \\ -\alpha_j \sinh(\alpha_j h) & \cosh(\alpha_j h) \end{bmatrix} \quad (j = l+1, l+2, \ldots l+m), \tag{2.100}$$

where

$$\begin{aligned} \kappa_i &= \left[k_0^2 n^2(x_i) - \beta^2\right]^{1/2} \\ \alpha_j &= \left[\beta^2 - k_0^2 n^2(x_j)\right]^{1/2}. \end{aligned} \tag{2.101}$$

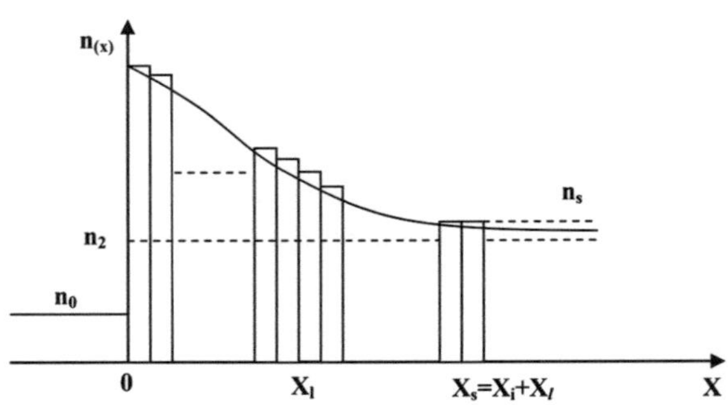

Fig. 2.6 Graded-index structure with only one turning point

According to the transfer matrix method, the corresponding matrix equation is as follows:

$$\begin{bmatrix} E_y(0) \\ E'_y(0) \end{bmatrix} = \left[\prod_{i=1}^{l} M_i\right]\left[\prod_{j=l+1}^{l+m} M_j\right]\begin{bmatrix} E_y(x_s) \\ E'_y(x_s) \end{bmatrix}, \tag{2.102}$$

while the evanescent behavior in both the outer claddings in the approximated multilayer structure are given by

$$E_y(x) = \begin{cases} A_0 \exp(p_0 x) & (x<0) \\ A_s \exp[-p_s(x-x_s)] & (x>x_s) \end{cases}, \tag{2.103}$$

where

$$\begin{aligned} p_0 &= \left(\beta^2 - k_0^2 n_0^2\right)^{1/2} \\ p_s &= \left(\beta^2 - k_0^2 n_s^2\right)^{1/2}. \end{aligned} \tag{2.104}$$

Substituting Eq. (2.103) into Eq. (2.102), we have

$$\begin{pmatrix} -p_0 & 1 \end{pmatrix}\left(\prod_{i=1}^{l} M_i\right)\left(\prod_{j=l+1}^{l+m} M_j\right)\begin{pmatrix} 1 \\ -p_s \end{pmatrix} = 0. \tag{2.105}$$

By a simple algebraic process, the above formula can be recast into

$$\begin{pmatrix} -p_0 & 1 \end{pmatrix}\left(\prod_{i=1}^{l} M_i\right)\begin{pmatrix} 1 \\ -p_{l+1} \end{pmatrix} = 0, \tag{2.106}$$

where

$$\begin{cases} p_j = \alpha_j \dfrac{\sinh(\alpha_j h) + \frac{p_{j+1}}{\alpha_j}\cosh(\alpha_j h)}{\cosh(\alpha_j h) + \frac{p_{j+1}}{\alpha_j}\sinh(\alpha_j h)} \\ \qquad (j = l+1, l+2, \ldots, l+m) \\ p_{l+m+1} = p_s \end{cases} \tag{2.107}$$

So it is clear that in Eq. (2.106), the field distribution outside the turning point is treated as a exponentially decaying field, that is,

$$E_y(x) = A_t \exp[-p_{l+1}(x-x_t)] \quad (x>x_t). \tag{2.108}$$

Similar with the process we used in the last section, the exact eigenvalue equation can be derived from Eq. (2.106)

$$\sum_{i=1}^{l}\kappa_i h+\Phi(s)=N\pi+\tan^{-1}\left(\frac{p_0}{\kappa_1}\right)+\tan^{-1}\left(\frac{p_{l+1}}{\kappa_l}\right),\quad (N=0,1,2,\ldots) \tag{2.109}$$

where

$$\Phi(s)=\sum_{i=1}^{l-1}\left[\Phi_{i+1}-\tan^{-1}\left(\frac{\kappa_{i+1}}{\kappa_i}\tan\Phi_{i+1}\right)\right], \tag{2.110}$$

$$\Phi_i=\tan^{-1}\left(\frac{p_i}{\kappa_i}\right), \tag{2.111}$$

$$p_i=\kappa_i\tan\left[\tan^{-1}\left(\frac{p_{i+1}}{\kappa_i}\right)-\kappa_i h\right]\quad (i=1,2,\ldots,1), \tag{2.112}$$

and $\tan^{-1}\left(\frac{p_{l+1}}{\kappa_l}\right)$ denotes the phase shift at the turning point.

Next considering the graded-index structure, i.e., $l\to\infty, m\to\infty$, the first term in the left-hand side of Eq. (2.109) becomes integral

$$\sum_{i=1}^{l}\kappa_i h_i=\int_0^{x_t}\kappa(x)\mathrm{d}x, \tag{2.113}$$

and the second term becomes

$$\sum_{i=1}^{l-1}\left[\Phi_{i+1}-\arctan\left(\frac{\kappa_{i+1}}{\kappa_i}\tan\Phi_{i+1}\right)\right]=\int_0^{x_y}\frac{q}{q^2+\kappa^2}\frac{\mathrm{d}\kappa}{\mathrm{d}x}\mathrm{d}x. \tag{2.114}$$

where $q(x)=-E'(x)/E(x)$. Finally, the exact eigenvalue equation of the graded-index slab waveguide can be written as follows:

$$\int_0^{x_t}\kappa dx+\Phi(s)=\int_0^{x_t}\left(\kappa+\frac{q}{q^2+\kappa^2}\frac{\mathrm{d}\kappa}{\mathrm{d}x}\right)\mathrm{d}x=N\pi+\tan^{-1}\left(\frac{p_0}{\kappa_1}\right)+\tan^{-1}\left(\frac{p_{l+1}}{\kappa_l}\right).\quad (N=0,1,2,\ldots) \tag{2.115}$$

It should be note that the term $\Phi(s)$ denotes the phase contribution of the scattered subwaves, and $\frac{q}{q^2+\kappa^2}\frac{\mathrm{d}\kappa}{\mathrm{d}x}$ represents its wave number. The above equation can be generalized to the case with two turning points without difficult. Consider the graded-index waveguide with two turning points at x_{t1} and x_{t2}, and extend to infinity on both sides (Fig. 2.7).

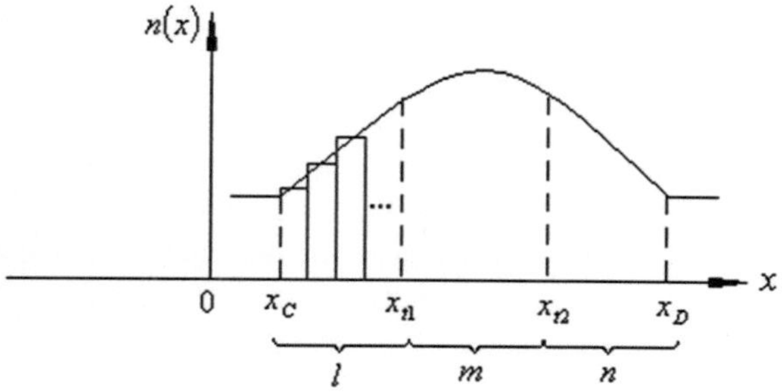

Fig. 2.7 A graded-index waveguide with two turning points

Take the similar process used above, first we truncate the index profile $n(x)$ at x_C and x_D, respectively, and divide the regions (x_C, x_{t1}), (x_{t1}, x_{t2}), and (x_{t2}, x_D) into l, m, and n segments of the same thickness h. Then, the dispersion equation via the transfer matrix can be written as follows:

$$\int_{x_{t1}}^{x_{t2}} \kappa \mathrm{d}x + \Phi(s) = N\pi + \tan^{-1}\left(\frac{p_l}{\kappa_{l+1}}\right) + \tan^{-1}\left(\frac{p_{l+m+1}}{\kappa_{l+m}}\right), \quad (N = 0, 1, 2, \ldots) \tag{2.116}$$

where

$$\begin{aligned} \Phi(s) &= \sum_{j=l+1}^{l+m-1}\left[\Phi_{j+1} - \tan^{-1}\left(\frac{\kappa_{j+1}}{\kappa_j}\tan\Phi_{j+1}\right)\right] \\ \Phi_j &= \tan^{-1}\left(\frac{p_j}{\kappa_j}\right). \end{aligned} \tag{2.117}$$

p_l and p_{l+m+1} are the effective attenuation coefficients for the regions ($x < x_{t1}$) and ($x > x_{t2}$), respectively, which are specified by

$$p_k = \alpha_k \frac{\sinh(\alpha_k h) + \frac{p_{k+1}}{\alpha_k}\cosh(\alpha_k h)}{\cosh(\alpha_k h) + \frac{p_{k+1}}{\alpha_k}\sinh(\alpha_k h)} \quad (k = l+m+1, l+m+2, \ldots, l+m+n), \tag{2.118}$$

where $p_0 = p_C$ and $p_C = \left[\beta^2 - k_0^2 n^2(x_c)\right]^{1/2}$,

$$p_i = \alpha_i \frac{\sinh(\alpha_i h) + \frac{p_{i-1}}{\alpha_i}\cosh(\alpha_i h)}{\cosh(\alpha_i h) + \frac{p_{i-1}}{\alpha_i}\sinh(\alpha_i h)}, \quad (i = 1, 2, \ldots, l) \tag{2.119}$$

where $p_{l+m+n+1} = p_D$ and $p_D = \left[\beta^2 - k_0^2 n^2(x_D)\right]^{1/2}$.

2.5.2 The Phase Shift at Turning Point [7]

From the last section, it is clear that we can replace the field distribution outside the turning point with an exponentially decaying field without introducing any calculation error. So it is possible to treat the index profile outside the turning point as with a constant n_{eq}, which is smaller than $n(x_t)$ (see Fig. 2.8).

According to the analysis above, the effective attenuation coefficient can be written as follows:

$$p_t = \left(\beta^2 - k_0^2 n_{\text{eq}}^2\right)^{1/2}, \tag{2.120}$$

If we restrict ourselves with the bound electromagnetic modes, p_t must be a finite and positive quantity. Let us prove this statement briefly below.

(1) For $j = l$+m, consider Eqs. (2.101), (2.104), and (2.107), and there is

$$p_c = p_{l+m+1} > \alpha_{l+m}; \tag{2.121}$$

since both α_j and h are positive real number, it follows

$$\cosh(\alpha_{l+m}h) > \sinh(\alpha_{l+m}h); \tag{2.122}$$

and then, in view of (2.107), one can prove that $p_{l+m} > \alpha_{l+m}$. Repeat the above process, and finally, there is

$$p_{l+1} > \alpha_{l+1}. \tag{2.123}$$

(2) Let us rewrite Eq. (2.107) into the following formation

$$p_j = p_{j+1} \frac{\cosh(\alpha_j h) + \dfrac{\alpha_j}{p_{j+1}} \sinh(\alpha_j h)}{\cosh(\alpha_j h) + \dfrac{p_{j+1}}{\alpha_j} \sinh(\alpha_j h)}. \tag{2.124}$$

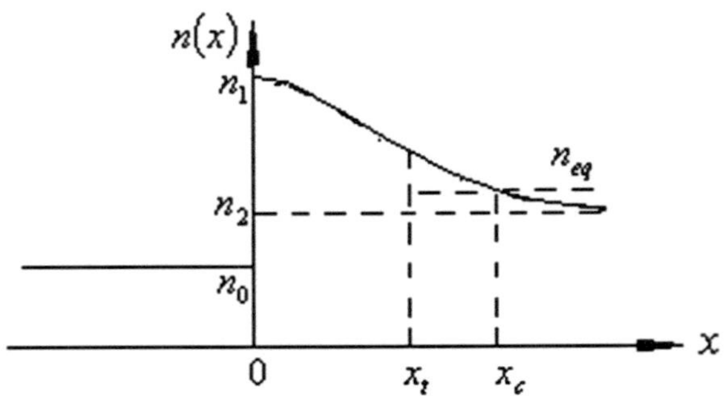

Fig. 2.8 Replace the index distribution outside the turning point with an effective refractive index n_{eq}

Obviously, $p_j < p_{j+1}$ holds, which leads to

$$p_t = p_{l+1} < p_{l+m+1} = p_c. \tag{2.125}$$

(3) According to Eqs. (2.123) and (2.125), one obtains

$$\alpha_{l+1} < p_t < p_c, \tag{2.126}$$

which shows that p_t is a finite and real number. However, according to Eq. (2.101), as $l \to \infty (h \to 0)$, there holds

$$\kappa_l = \left[k_0^2 n^2(x_l) - \beta^2\right]^{1/2} \to \left[k_0^2 n^2(x_t) - \beta^2\right]^{1/2} = 0. \tag{2.127}$$

Finally, the phase shift at the turning point can be calculated as follows:

$$\tan^{-1}\left(\frac{p_t}{k_l}\right) = \tan^{-1}\sqrt{\frac{\beta^2 - k_0^2 n_{\text{eq}}^2}{k_0^2 n^2(x_l) - \beta^2}} \to \frac{\pi}{2} \quad (l \to \infty), \tag{2.128}$$

which is exact twice the result in the basic WKB approximation. In the modified WKB approximation, the non-integral Maslov index is used, which allows the reflection phase at the turning points approaches $\pi/2$ in the semiclassical limit, and approaches π in the anticlassical limit. Compared with the WKB approximation, the result we derived has the following features:

(1) The phase shift at the turning point is constant π, which is independent of the propagation constant and the refractive index distribution; it is a general result.
(2) The phase shift does not related to the position of the turning points, and it does not vary if the turning point is near truncated points, discontinuous points, or other turning points.
(3) The phase shift is the same for different wavelengths.

Using Eq. (2.128), we can further simplify the eigenvalue equation Eq. (2.115) of an arbitrary graded-index waveguide with only one turning point as plotted in Fig. 2.6 as

$$\int_0^{x_t} \kappa \mathrm{d}x + \Phi(s) = m\pi + \tan\left(\frac{p_0}{\kappa_1}\right) + \frac{\pi}{2} \quad (m = 0, 1, 2, \ldots), \tag{2.129}$$

and the eigenvalue equation of graded-index waveguide with two turning points in Fig. 2.7 as

$$\int_{x_{t1}}^{x_{t2}} k \mathrm{d}x + \Phi(s) = (m+1)\pi \quad (m = 0, 1, 2, \ldots). \tag{2.130}$$

In conclusion of this section, we started with the transfer matrix and applied to multilayer and graded-index waveguide to obtain an exact and general eigenvalue equation with clear physics. The notion of scattered subwaves, which is completely neglected in semiclassical theories, was proposed and appeared in all the obtained eigenvalue equations.

References

1. M. Born, W. Wolf, Principles of Optics, 6th edn (corrected) (Pergamon Press, Oxford, 1986)
2. A. Yariv, Quantum Electronics, 2nd edn (Wiley, New York 1975)
3. L.I. Schiff, Quantum Mechanics (McGraw-Hill, New York, 1955)
4. A. Gedeon, Comparison between rigorous theory and WKB-analysis of modes in graded-index waveguides. Opt. Commun. **12**, 329 (1974)
5. H. Friedrich, J. Trost, Working with WKB waves far from the semiclassical limit. Phys. Rep. **397**, 359 (2004)
6. M.J. Sun, M.W. Muller, W.S.C. Chang, Thin-film waveguide gyrators: a theoretical analysis. Appl. Opt. **16**, 2986 (1977)
7. Z. Cao, Q. Liu, Y. Jiang, Q. Shen, X. Dou, Y. Ozaki, Phase shift at a turning point in a planar optical waveguide. J. Opt. Soc. Am. A **18**, 216 (2001)

Chapter 3
Periodic Waveguides and MQW Waveguide

Abstract In research field of optics, diffraction gratings are widely investigated and applied; on the other hand, periodic metal waveguide is familiar to those worked with microwave waveguide. The early studies on periodic waveguide have a profound impact on the development of the waveguide optics. Currently, periodic waveguide has become an important integrated optical element and obtains a wide range of applications in functional devices such as the grating couplers, filters, distributed feedback lasers, and the distributed Bragg reflector lasers. The most common analysis method on the periodic waveguide is the coupled mode theory (Yariv in Quantum electronics. Wiley, New York, 1975), which is extremely effective to deal with the coupling between different modes. However, the method is cumbersome in mathematics and lack of clearness in giving physical insight. In this chapter, we introduce another method (viz. the transfer matrix method) to analyze the periodic waveguide. The simple rectangular periodic waveguide is discussed first, and then, the periodic waveguide of arbitrary shape is also studied.

Keywords Periodic waveguides · Transfer matrix · Coupling coefficient · Multiple quantum well optical waveguide

3.1 Rectangular Corrugated Periodic Waveguide [2]

A simple method to study the rectangular periodic waveguide is presented in this section. It is able to present exact analytical expressions with a clear physical picture.

3.1.1 From Corrugated Optical Waveguide to Rectangular Periodic Waveguide

Rectangular periodic waveguide is shown in Fig. 3.1, where the refractive index of the guiding layer, substrate, and the covering layer is denoted by n_1, n_2, and n_3,

X. Wang et al., *Progress in Planar Optical Waveguides*,
Springer Tracts in Modern Physics 266, DOI 10.1007/978-3-662-48984-0_3

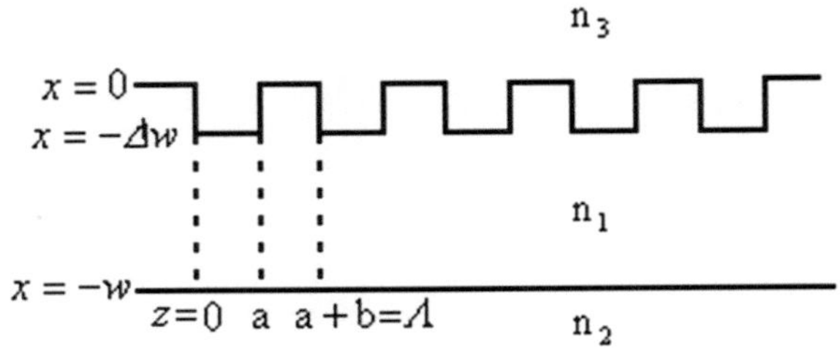

Fig. 3.1 Rectangular periodic waveguide

respectively. Light travels along the z direction. Characters a and b in Fig. 3.1 are the width of the corrugated well and the barrier regions, respectively. $\Lambda = a + b$ is the period of corrugated structure. w is the thickness of the guiding layer, while Δw is the height of the corrugated structure. Usually, the waveguide with the corrugated structure is called as the perturbed waveguide, and those without the corrugated structure are named as the unperturbed waveguide. For simplicity, this chapter only discusses the TE mode. The TM mode is left for the readers.

The eigenmode equation of TE polarization in the unperturbed waveguide is

$$k_x w = m\pi + \tan^{-1}\left(\frac{q}{k_x}\right) + \tan^{-1}\left(\frac{p}{k_x}\right), \quad (m = 0, 1, 2, \ldots), \tag{3.1}$$

where

$$\left.\begin{aligned} q &= \left(k_z^2 - k_0^2 n_3^2\right)^{1/2} \\ p &= \left(k_z^2 - k_0^2 n_2^2\right)^{1/2} \\ k_x &= \left(k_0^2 n_1^2 - k_z^2\right)^{1/2} \end{aligned}\right\}, \tag{3.2}$$

k_z is the propagation constant of the guided modes. Using the relation between k_x and k_z, the expressions for q and p can be expressed as functions of k_x, namely

$$\left.\begin{aligned} q &= \left[k_0^2\left(n_1^2 - n_3^2\right) - k_x^2\right]^{1/2} \\ p &= \left[k_0^2\left(n_1^2 - n_2^2\right) - k_x^2\right]^{1/2} \end{aligned}\right\}. \tag{3.3}$$

As a result, if the refractive index of the three layers in the waveguide and the wavelength is fixed, Eq. (3.1) depends only on the k_x and w. Solving Eq. (3.1), $\mathrm{d}k_x/\mathrm{d}w$ can be written as

$$\frac{\mathrm{d}k_x}{k_x} = -\frac{\mathrm{d}w}{w_{\mathrm{eff}}}, \tag{3.4}$$

where the effective thickness of the waveguide is

$$w_{\mathrm{eff}} = w + \frac{1}{q} + \frac{1}{p}. \tag{3.5}$$

In case that $\Delta w \ll w$, perturbation method can be applied to simplify Eq. (3.4) approximately as

$$\frac{\Delta k_x}{k_x} = -\frac{\Delta w}{w_{\mathrm{eff}}}. \tag{3.6}$$

From the relation between k_x and k_z, we have

$$k_x \Delta k_x = -k_z \Delta k_z. \tag{3.7}$$

Substituting Eq. (3.7) into (3.6), it yields

$$\frac{\Delta k_z}{k_z} = \left(\frac{k_x}{k_z}\right)^2 \frac{\Delta w}{w_{\mathrm{eff}}}. \tag{3.8}$$

Equation (3.8) indicates the eigenvalue of the guided mode will vary if the guiding layer thickness changes. We can rewrite Eq. (3.8) in the notion of effective refractive index as follows:

$$\frac{\Delta N}{\Delta w} = \frac{N}{w_{\mathrm{eff}}} \left(\frac{n_1^2}{N^2} - 1\right), \tag{3.9}$$

where $N = k_z/k_0$. Since the effective refractive index of the waveguide ranges between the substrate and the guiding layer, i.e., $n_2 < N < n_1$, it can be seen from Eq. (3.9) that

$$\frac{\Delta N}{\Delta w} > 0, \tag{3.10}$$

which shows that the effective refractive index N of the waveguide is a monotone increasing function of the guiding layer thickness w. So in the rectangular corrugated periodic waveguide plotted in Fig. 3.1, it is reasonable to argue that those regions of the corrugated barriers have a larger effective refractive index than the rest regions, and the resulted effective refractive index of a periodic perturbed waveguide is a periodic function of the space. In other words, it is possible to treat the rectangular corrugated periodic waveguide in Fig. 3.1 as an optical multilayer film structure as plotted in Fig. 3.2 whose refractive index is also periodic in space. The above discussion that turning the two-dimensional electromagnetic problem into a simple one-dimensional problem lays the foundation for further studying the propagation characteristics of the periodic waveguide.

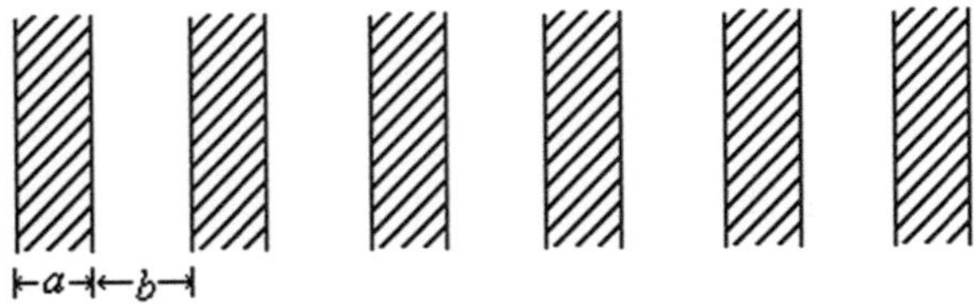

Fig. 3.2 Optical multilayer film with periodic refractive index

3.1.2 Transfer Matrix and the Coupling Coefficient

The planar wave travels in the optical periodic multilayer film satisfies the following wave equation

$$\frac{\partial^2 E_y}{\partial z^2} + k_z^2 E_y = 0. \tag{3.11}$$

The propagation constant of the optical film corresponding to the corrugated well regions in the waveguide is $k_z = \alpha$, and that for the corrugated barrier regions in the waveguide is $k_z = \beta$. Form Eq. (3.10), it is obvious that $\beta > \alpha$, and let us define

$$\alpha = \beta - \Delta\beta. \tag{3.12}$$

Since $\Delta w \ll w$, there is

$$\Delta\beta \ll \beta. \tag{3.13}$$

In the corrugated well regions $0<z<a$ as shown in Fig. 3.1, the solution to Eq. (3.11) can be written as

$$E_y(z) = Ce^{i\alpha z} + De^{-i\alpha z}. \tag{3.14}$$

The first-order derivative of $E_y\,(z)$ on z is

$$E_y'(z) = i\alpha(Ce^{i\alpha z} - De^{-i\alpha z}). \tag{3.15}$$

Inserting $z = 0$ and $z = a$ into Eqs. (3.14) and (3.15), we can get the matrix equation

$$\begin{bmatrix} E_y(a) \\ E_y'(a) \end{bmatrix} = \begin{bmatrix} \cos(\alpha a) & \frac{1}{\alpha}\sin(\alpha a) \\ -\alpha\,\sin(\alpha a) & \cos(\alpha a) \end{bmatrix} \begin{bmatrix} E_y(0) \\ E_y'(0) \end{bmatrix}. \tag{3.16}$$

By defining

$$M(a) = \begin{bmatrix} \cos(\alpha a) & \frac{1}{\alpha}\sin(\alpha a) \\ -\alpha \sin(\alpha a) & \cos(\alpha a) \end{bmatrix}, \tag{3.17}$$

it is clear that the matrix $M(a)$ is mathematically equivalent to the transfer matrix introduced in Chap. 2. But in this case, the transfer matrix connects the field and its first derivative of the two boundaries in the longitudinal direction (in the z direction). So the application range of the transfer matrix is very broad.

Using the same step, the corresponding matrix of the corrugated barrier region ($a<z<a+b$) is

$$M(b) = \begin{bmatrix} \cos(\beta b) & \frac{1}{\beta}\sin(\beta b) \\ -\beta \sin(\beta b) & \cos(\beta \mathrm{b}) \end{bmatrix}. \tag{3.18}$$

Using the characteristics of the transfer matrix, it is clear that the matrix represents the perturbation period ($\Lambda = a+b$) is the product of the two matrix $M(a)$ and $M(b)$, namely

$$M(\Lambda) = M(b) \cdot M(a). \tag{3.19}$$

The multiplication order of Eq. (3.19) cannot be varied, which indicates the propagation direction of the light. The propagation direction defined in Eq. (3.19) is along the negative z-axis direction, and if the light travels in the positive z-axis direction, the two matrixes in the right side of Eq. (3.19) should be replaced by their inverse matrix and the multiplication order should be changed.

Combing Eqs. (3.17), (3.18), and (3.19), it is not difficult to derive the transfer matrix for a perturbation period

$$M(\Lambda) = \begin{bmatrix} m_{11} & m_{12} \\ m_{21} & m_{22} \end{bmatrix}, \tag{3.20}$$

where

$$\left.\begin{aligned} m_{11} &= \cos(\beta b)\cos(\alpha a) - \frac{\alpha}{\beta}\sin(\beta b)\sin(\alpha a) \\ m_{12} &= \frac{1}{\alpha}\cos(\beta b)\sin(\alpha a) + \frac{1}{\beta}\sin(\beta b)\cos(\alpha a) \\ m_{21} &= -\beta \sin(\beta b)\cos(\alpha a) - \alpha \cos(\beta b)\sin(\alpha a) \\ m_{22} &= \cos(\beta b)\cos(\alpha a) - \frac{\beta}{\alpha}\sin(\beta b)\sin(\alpha a) \end{aligned}\right\}. \tag{3.21}$$

Assume the periodic waveguide has n perturbation period, and the matrix is the same for all the perturbation periods, as a result the whole waveguide can be represented by n th power of a single matrix, namely

$$M(n\Lambda) = \underbrace{M(\Lambda) \cdot M(\Lambda) \ldots M(\Lambda)}_{n} = [M(\Lambda)]^n. \tag{3.22}$$

According to the characteristics of the transfer matrix, there is

$$M(n\Lambda) = \begin{bmatrix} m_{11}U_{n-1}(\chi) - U_{n-2}(\chi) & m_{12}U_{n-1}(\chi) \\ m_{21}U_{n-1}(\chi) & m_{22}U_{n-1}(\chi) - U_{n-2}(\chi) \end{bmatrix}, \tag{3.23}$$

where

$$\chi = \frac{1}{2}(m_{11} + m_{22}), \tag{3.24}$$

and $U_n(\chi)$ is the second kind of the Chebyshev polynomial

$$U_n(\chi) = \frac{\sin[(n+1)\cos^{-1}\chi]}{\sqrt{1-\chi^2}}. \tag{3.25}$$

In order to facilitate the later calculation, we define

$$M(n\Lambda) = \begin{bmatrix} \mu_{11} & \mu_{12} \\ \mu_{21} & \mu_{22} \end{bmatrix}, \tag{3.26}$$

and based on Eqs. (3.21) and (3.24), it is easy to derive

$$\chi = \cos(\beta b)\cos(\alpha a) - \frac{1}{2}\left(\frac{\beta}{\alpha} + \frac{\alpha}{\beta}\right)\sin(\beta b)\sin(\alpha a). \tag{3.27}$$

In view of Eqs. (3.12) and (3.13), treating $\Delta\beta/\beta$ as first-order infinitesimal, and Eq. (3.27) can be approximated as the following equation when the second-order infinitesimal is reserved.

$$\chi = \cos(\beta b + \alpha a) - \frac{1}{2}\left(\frac{\Delta\beta}{\beta}\right)^2 \sin(\beta b)\sin(\alpha a). \tag{3.28}$$

Setting

$$\chi = \cos(\beta' \Lambda), \tag{3.29}$$

And according to the physical meaning of β and α, it is clear that β' is the propagation constant in the equivalent periodic optical waveguide and the Bloch wave vector of the matter wave in the periodic perturbed lattice. When the optical period of the perturbation period equals an odd times of a quarter of wavelength, two cases are of particular importance:

1. When the thickness of the corrugated barrier equals that of the corrugated well, equivalently

$$\alpha a = \beta b \approx (2l+1)\frac{\pi}{2}, \quad (l = 0, 1, 2, \ldots). \tag{3.30}$$

2. When the difference between these two parts equals a half of wavelength, for example,

$$\alpha a \approx (2l+1)\frac{\pi}{2}, \text{ and } \beta b = (2l+3)\frac{\pi}{2}. \tag{3.31}$$

Apparently, in both cases there are

$$\alpha a + \beta b \approx m\pi, \quad (m = 0, 1, 2, \ldots). \tag{3.32}$$

Considering Eq. (3.28), we can find $|\chi|$ is larger than 1, and following the definition of Eq. (3.29), it can be shown that the propagation constant β' in the periodic perturbed region is complex. Assume β' satisfies the following relation

$$\beta' = \beta_0 + is, \tag{3.33}$$

where $\beta_0 = m\pi/\Lambda$ and s, which is the attenuation coefficient, is a real number. If we call $\alpha a + \beta b = m\pi$ as the phase-matching condition, then its minor deviation can be defined as

$$\delta = m\pi - (\alpha a + \beta b). \tag{3.34}$$

Using Eqs. (3.29) and (3.33), we can have

$$\begin{aligned} \chi &= \cos(\beta'\Lambda) \\ &= \cos(m\pi + is\Lambda) \\ &= (-1)^m \cosh(s\Lambda), \end{aligned} \tag{3.35}$$

and

$$\cos(\beta b + \alpha a) = \cos(m\pi - \delta) = (-1)^m \cos\delta. \tag{3.36}$$

Checking the second term on the right side of Eq. (3.28), when Eq. (3.30) holds (m is odd), the product of two sine functions is positive; when Eq. (3.31) holds (m is even), the product of the two sine functions is negative. In both cases, Eq. (3.28) becomes

$$\cosh(s\Lambda) = \cos\,\delta + \frac{1}{2}\left(\frac{\Delta\beta}{\beta}\right)^2. \quad (3.37)$$

When s and δ are sufficiently small, applying the series expansion method and ignoring all the infinitesimal higher than two orders, Eq. (3.37) can be simplified as

$$s^2 = \left(\frac{\Delta\beta}{\beta\Lambda}\right)^2 - \left(\frac{\delta}{\Lambda}\right)^2. \quad (3.38)$$

Define

$$\kappa_c = \frac{\Delta\beta}{\beta\Lambda}, \quad (3.39)$$

which denotes the relative variation of the propagation constant in a perturbation period. Comparing with the relative parameter in the coupled wave analysis [1], it shows that κ_c is the coupling coefficient. The magnitude of the coupling coefficient reflects the energy exchanging rate between the "forward traveling wave" and the "backward traveling wave."

It can be drawn from Eq. (3.8) that

$$\frac{\Delta\beta}{\beta} = \frac{\kappa^2\Delta w}{\beta^2 w_{\mathrm{eff}}}, \quad (3.40)$$

where the k_x and k_z have been replaced by κ and β, and the coupling coefficient expression can be obtained via inserting Eq. (3.40) into (3.39)

$$\kappa_c = \frac{\kappa^2\Delta w}{\beta^2\Lambda w_{\mathrm{eff}}}. \quad (3.41)$$

The physics behind the above coupling coefficient is rather clear: The magnitude of the κ_c is proportional to the perturbation Δw and the square of the transverse wave vector κ^2 and inversely proportional to the perturbation period Λ, effective thickness w_{eff} of the waveguide, and the square of the longitudinal wave vector β^2. In the coupled wave analysis, the coupling coefficient is expressed via the integral of the transverse mode, and it is far more complicate than Eq. (3.41). In the next section, we will see that the result obtained by the transfer matrix method is more accurate. Moreover, the physical meaning of the approximation condition $\Delta w \ll w$ applied in this section is clearer than the approximation condition used in the coupled wave analysis.

3.1.3 Forward and Backward Traveling Waves

In this section, the corresponding matrix $M(n\Lambda)$ of the whole perturbed periodic waveguide is derived, and the explicit expressions of the forward and backward traveling waves are also derived from the boundary conditions of the perturbed region. Finally, the transmission and reflectivity of the periodic waveguide are obtained. For clarity, in this section, we only consider the case of $m = 1$.

Combining Eqs. (3.25) and (3.33), we can have

$$\begin{aligned} U_{n-1}(\chi) &= \frac{\sin(n\cos^{-1}\chi)}{\sqrt{1-\chi^2}} \\ &= (-1)^n \frac{\sinh(ns\Lambda)}{\sinh(s\Lambda)}, \end{aligned} \tag{3.42}$$

Considering Eqs. (3.12), (3.13), and (3.21), it follows

$$\left.\begin{aligned} m_{11} &= \cos(\beta b+\alpha a)+\frac{\Delta\beta}{\beta}\sin\beta b\sin\alpha a \\ m_{12} &= \frac{1}{\beta}\sin(\beta b+\alpha a)+\frac{\Delta\beta}{\beta^2}\cos\beta b\sin\alpha a \\ m_{21} &= -\beta\sin(\beta b+\alpha a)+\Delta\beta\cos\beta b\sin\alpha a \\ m_{22} &= \cos(\beta b+\alpha a)-\left[\frac{\Delta\beta}{\beta}+\left(\frac{\Delta\beta}{\beta}\right)^2\right]\sin\beta b\sin\alpha a \end{aligned}\right\}. \tag{3.43}$$

Then, the below expressions can be derived via Eqs. (3.28), (3.29), (3.32), (3.33), and (3.34)

$$\left.\begin{aligned} \cos(\beta b+\alpha a) &= -\cosh(s\Lambda)+\frac{1}{2}\left(\frac{\Delta\beta}{\beta}\right)^2\sin\beta b\sin\alpha a \\ \sin(\beta b+\alpha a) &\approx \delta \end{aligned}\right\}. \tag{3.44}$$

Inserting Eq. (3.44) into (3.43), and using Eq. (3.23), the corresponding matrix of whole perturbed periodic waveguide can be written as

$$M(n\Lambda) = \begin{bmatrix} (-1)^{n-1}\cosh(ns\Lambda)+\Delta_{11} & (-1)^n\frac{\delta}{\beta s\Lambda}\sinh(ns\Lambda)+\Delta_{12} \\ (-1)^{n-1}\frac{\beta\delta}{s\Lambda}\sinh(ns\Lambda)+\Delta_{21} & (-1)^{n-1}\cosh(ns\Lambda)+\Delta_{22} \end{bmatrix} \tag{3.45}$$

where

$$\Delta_{11} = (-1)^n \frac{\sinh(ns\Lambda)}{\sinh(s\Lambda)} \cdot \left[\frac{\Delta\beta}{\beta} + \frac{1}{2}\left(\frac{\Delta\beta}{\beta}\right)^2\right] \sin \beta b \sin \alpha a,$$
$$\Delta_{12} = (-1)^n \frac{\sinh(ns\Lambda)}{\sinh(s\Lambda)} \cdot \frac{\Delta\beta}{\beta^2} \cos \beta b \sin \alpha a,$$
$$\Delta_{21} = (-1)^n \frac{\sinh(ns\Lambda)}{\sinh(s\Lambda)} \cdot \Delta\beta \cos \beta b \sin \alpha a,$$
$$\Delta_{22} = (-1)^{n-1} \frac{\sinh(ns\Lambda)}{\sinh(s\Lambda)} \cdot \left[\frac{\Delta\beta}{\beta} + \frac{1}{2}\left(\frac{\Delta\beta}{\beta}\right)^2\right] \sin \beta b \sin \alpha a.$$

Although the above quantities are small when compared with the main parts of the matrix elements, they cannot be ignored in the calculation.

The next step is to solve the exact expression of the forward and backward traveling waves based on the transfer matrix $M(n\Lambda)$ of the whole perturbed periodic waveguide. For this aim, the boundary condition must be determined. Let us assume the amplitude of the incident wave on the boundary $z = 0$ is 1, and the reflected wave has an amplitude of A, while the amplitude of the transmitted wave on the boundary of $z = n\Lambda$ is denoted by B. According to Fig. 3.3, the incident, reflected, and transmitted waves can be related by

$$B\begin{bmatrix} 1 \\ i\beta \end{bmatrix} e^{i\beta n\Lambda} = \begin{bmatrix} \mu_{11} & \mu_{12} \\ \mu_{21} & \mu_{22} \end{bmatrix} \begin{bmatrix} 1 + A \\ i\beta(1 - A) \end{bmatrix}. \tag{3.46}$$

It is not difficult via Eq. (3.46) to find the incident wave at the boundary of $z = n\Lambda$

$$\begin{aligned} Be^{i\beta n\Lambda} &= \frac{2i\beta}{(\beta^2\mu_{12} - \mu_{21}) + i\beta(\mu_{11} + \mu_{22})} \\ &= \frac{is\, e^{i\beta_0 n\Lambda}}{\frac{\delta}{\Lambda}\sinh(ns\Lambda) - is\cosh(ns\Lambda)}, \end{aligned} \tag{3.47}$$

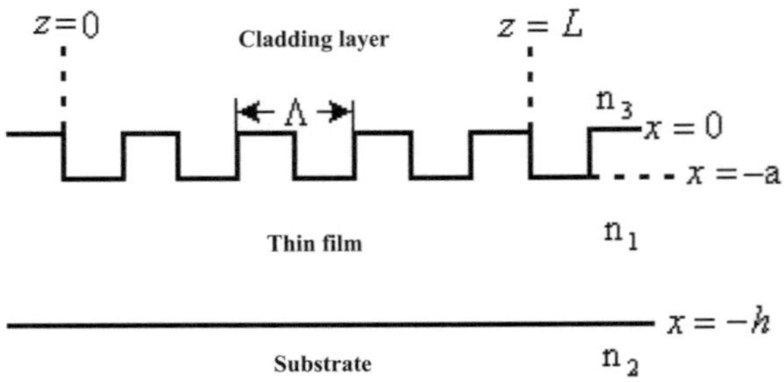

Fig. 3.3 The structure of a periodic waveguide

And the reflected wave at the boundary $z = 0$ is

$$
\begin{aligned}
A &= \frac{(\mu_{21} + \beta^2 \mu_{12}) + i\beta(\mu_{22} - \mu_{11})}{(\beta^2 \mu_{12} - \mu_{21}) + i\beta(\mu_{11} + \mu_{22})} \\
&= \frac{i\kappa_c \sinh(ns\Lambda)}{\frac{\delta}{\Lambda}\sinh(ns\Lambda) - is\cosh(ns\Lambda)}.
\end{aligned}
\tag{3.48}
$$

The two equations above are the same with the corresponding results derived from coupled wave analysis.

The transmission probability can be derived from Eq. (3.47)

$$
\begin{aligned}
T &= BB^* \\
&= \frac{s^2}{\left(\frac{\delta}{\Lambda}\right)^2 \sinh^2(ns\Lambda) + s^2\cosh^2(ns\Lambda)}.
\end{aligned}
\tag{3.49}
$$

Similarly, the reflectivity can be derived via Eq. (3.48):

$$
\begin{aligned}
R &= AA^* \\
&= \frac{\kappa_c^2 \sinh^2(ns\Lambda)}{\left(\frac{\delta}{\Lambda}\right)^2 \sinh^2(ns\Lambda) + s^2\cosh^2(ns\Lambda)}.
\end{aligned}
\tag{3.50}
$$

If we calculate the sum of Eqs. (3.49) and (3.50), and using the following relation

$$
\kappa_c^2 = s^2 + \left(\frac{\delta}{\Lambda}\right)^2,
\tag{3.51}
$$

it is easy to find that

$$
T + R = 1.
\tag{3.52}
$$

This demonstrates the physical significance of the fact that the transfer matrix is a unit matrix, and in other words, this characteristic of the transfer matrix reflects the conservation of energy.

Combing Eqs. (3.34) and (3.40), we have

$$
\frac{\delta}{\Lambda} = \frac{\kappa^2 \Delta w}{2\beta w_{\text{eff}}} - (\beta - \beta_0),
\tag{3.53}
$$

and from Eq. (3.38), the condition for s to take its maximum is

$$
\frac{\delta}{\Lambda} = 0.
\tag{3.54}
$$

In this case $s = \kappa_c$, so the corresponding value of the perturbation period based on Eq. (3.53) is found to be

$$\begin{aligned}\Lambda &= \pi\left(\beta - \frac{\kappa^2 \Delta w}{2\beta w_{\text{eff}}}\right)^{-1} \\ &\approx \frac{\pi}{\beta} + \frac{\pi\kappa^2 \Delta w}{2\beta^3 w_{\text{eff}}},\end{aligned} \tag{3.55}$$

For a fixed mode, when Eq. (3.55) is satisfied, the reflectivity of the waveguide reaches its maximum.

The condition for $s = 0$ is

$$\kappa_c = \pm\frac{\delta}{\Lambda}, \tag{3.56}$$

and this corresponds to the threshold value of the perturbation period

$$\Lambda \approx \frac{\pi}{\beta}\left[1 + \frac{(\pi \pm 2)\kappa^2 \Delta w}{2\beta^3 \Lambda w_{\text{eff}}}\right]. \tag{3.57}$$

The threshold value means that for a fixed mode, no observable reflected waves occur if the perturbation period Λ is smaller than the threshold value. In conclusion, if s is not zero, the allowable range of the perturbation period is

$$\frac{\pi}{\beta} + \frac{(\pi - 2)\kappa^2 \Delta w}{2\beta^3 w_{\text{eff}}} < \Lambda < \frac{\pi}{\beta} + \frac{(\pi + 2)\kappa^2 \Delta w}{2\beta^3 w_{\text{eff}}}. \tag{3.58}$$

The $\beta\Lambda$ in the denominator of the second terms on both sides of Eq. (3.58) is replaced by π, and since these two terms are first-order infinitesimal, such replacement is allowed.

Since $\Lambda = \pi/\beta_0$, the forbidden range of the propagation constant is

$$-\frac{\kappa^2 \Delta w}{\pi\beta w_{\text{eff}}} < \beta - \left(\beta_0 + \frac{\kappa^2 \Delta w}{2\beta w_{\text{eff}}}\right) < \frac{\kappa^2 \Delta w}{\pi\beta w_{\text{eff}}}, \tag{3.59}$$

or, equivalently

$$-\kappa_c < \beta - \left(\beta_0 + \frac{\kappa^2 \Delta w}{2\beta w_{\text{eff}}}\right) < \kappa_c. \tag{3.60}$$

Here, we introduce the notion of effective refractive index and set

$$\frac{\omega_0}{c} N = \beta_0 + \frac{\kappa^2 \Delta w}{2\beta w_{\text{eff}}}, \tag{3.61}$$

so

$$\frac{\omega}{c}N \approx \beta, \tag{3.62}$$

where the operator "≈" has two meanings: (1) the dispersion of the optical waveguide is nonlinear; (2) if ω and ω_0 are close, the effective refractive index can be approximated as constant, so we have

$$-\kappa_c < \frac{\omega - \omega_0}{c}N < \kappa_c. \tag{3.63}$$

From Eqs. (3.33), (3.38), (3.39), and (3.53), the propagation constant of the periodic corrugated waveguide can be expressed as

$$\begin{aligned}\beta' &= \beta_0 \pm is \\ &= \beta_0 \pm i\left[\kappa_c^2 - \left(\frac{N}{c}\right)^2(\omega - \omega_0)^2\right]^{1/2}.\end{aligned} \tag{3.64}$$

For later convenience, we multiply both sides of Eq. (3.64) by Λ, which yields

$$\beta'\Lambda = \pi \pm i\left[(\kappa_c\Lambda)^2 - \left(\Lambda\frac{N}{c}\right)^2(\omega - \omega_0)^2\right]^{1/2}. \tag{3.65}$$

When β' is real, the corresponding mode suffers no reflection attenuation, and there is

$$\Delta w = |\omega - \omega_0| \geq \frac{c\kappa_c}{N}. \tag{3.66}$$

It is not difficult to see that the forbidden range of the frequency is

$$\omega_0 - \frac{c\kappa_c}{N} < \omega < \omega_0 + \frac{c\kappa_c}{N}, \tag{3.67}$$

so the width of the forbidden region is

$$(\Delta\omega)_{\text{gap}} = \frac{2c\kappa_c}{N}. \tag{3.68}$$

β' is complex in the forbidden region, and based on Eq. (3.65), one can write

$$\text{Im}(\beta'\Lambda) = \left[(\kappa_c\Lambda)^2 - \left(\Lambda\frac{N}{c}\right)^2(\omega - \omega_0)^2\right]. \tag{3.69}$$

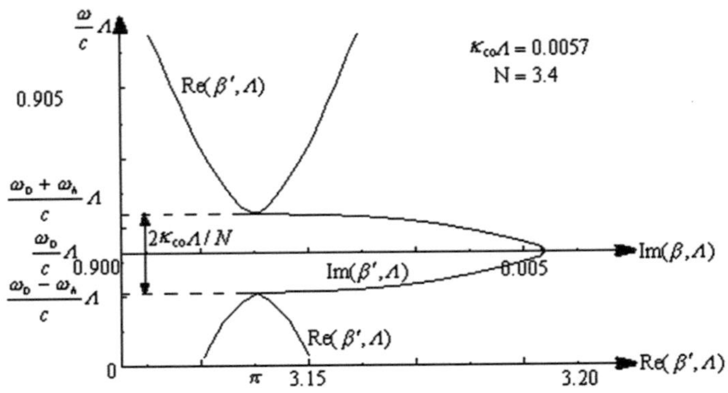

Fig. 3.4 The real and imaginary components of propagation constant β' versus frequency for the periodic waveguide

When $\omega = \omega_0$, it is obvious that

$$[\mathrm{Im}(\beta'\Lambda)]_{\max} = \kappa_c\Lambda. \tag{3.70}$$

This shows when $\omega = \omega_0$, the maximum of attenuation coefficient s equals the coupling coefficient κ_c. Consequently, a short section of the periodic waveguide acts as a highly reflected mirror of the wavelength near the Bragg frequency ω_0, and the maximum of the reflectivity according to Eq. (3.50) is

$$R_{\max} = [\tanh(\kappa_c L)]^2, \tag{3.71}$$

where L is the length of the perturbed region. Obviously, $R_{\max} \approx 1$ when $\kappa_c L \gg 1$.

The curves of the real component $\mathrm{Re}(\beta'\Lambda)$ and the imaginary component $\mathrm{Im}(\beta'\Lambda)$ as a function of ω in case of $m = 1$ are plotted in Fig. 3.4.

3.2 Corrugated Periodic Waveguide of Arbitrary Shape [3]

The above section has investigated the rectangular corrugated periodic waveguide via the transfer matrix method and obtained the analytical expression for the distributed feedback coefficient (or the coupling coefficient). In this section, we will extend to the corrugated periodic waveguide of arbitrary shape.

3.2.1 Analytical Expression for Coupling Coefficient

In the distributed feedback semiconductor laser, instead of using reflected mirror, the optical feedback is provided by the corrugated periodic waveguide. This is the reason why the coupling coefficient is also called as the distributed feedback coefficient. Since there are numerous kinds of the corrugated periodic waveguide, the primary object is to solve the distributed feedback coefficient of the periodic waveguide of arbitrary shape.

Consider the three-layer periodic waveguide in the distributed feedback laser as plotted in Fig. 3.5. For the sake of simplicity, we define the z-axis properly to let the corrugated shape function $f(z)$ satisfy the following equation

$$\int_{0}^{\Lambda} f(z)\mathrm{d}z = 0, \tag{3.72}$$

where Λ is the perturbation period.

For the corresponding unperturbed waveguide, its upside boundary is represented by z-axis. Set the thickness of the middle film of the unperturbed waveguide to be w, and the highest point and lowest point of the corrugated region are denoted by $x = g_2$ and $x = -g_1$, respectively. And the following equation holds

$$g_1 + g_2 = g \ll w. \tag{3.73}$$

Let us divide a perturbation period Λ into l parts and set the width of each region as a. For the unperturbed waveguide of thickness w, the propagation constant of the TE mode is β, and there is $\beta\Lambda \approx m\pi$, where m is an integer. In this section, we only focus on the case of $m = 1$. Based on the previous section, the transfer matrix of the ith part is given by

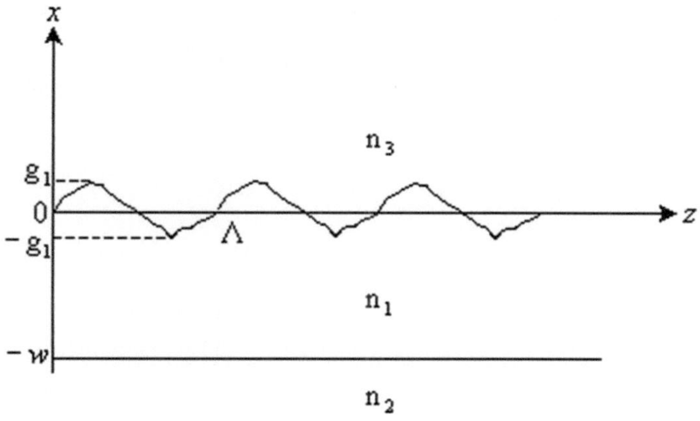

Fig. 3.5 Three-layer periodic waveguide in the distributed feedback laser

$$M_i = \begin{bmatrix} \cos(\beta+\Delta\beta_i)a & \frac{1}{\beta+\Delta\beta_i}\sin(\beta+\Delta\beta_i)\mathrm{a} \\ -(\beta+\Delta\beta_i)\sin(\beta+\Delta\beta_i)a & \cos(\beta+\Delta\beta_i)a \end{bmatrix}, \tag{3.74}$$

where

$$\Delta\beta_i = \frac{\kappa^2 f(z_i)}{\beta w_{\mathrm{eff}}}, \quad (i = 1, 2, \ldots, l). \tag{3.75}$$

From Eq. (3.73), it is clear that $\Delta\beta_i \ll \beta$.

Expanding the matrix M_i in the series of $\Delta\beta_i a$ leads to the follow expression

$$\begin{aligned} M_i = & \begin{bmatrix} \cos(\beta a) & \frac{1}{\beta}\sin(\beta \mathrm{a}) \\ -\beta\sin(\beta a) & \cos(\beta \mathrm{a}) \end{bmatrix} \\ & + (\Delta\beta_i a)\begin{bmatrix} -\sin(\beta a) & \frac{1}{\beta}\cos(\beta \mathrm{a}) - \frac{1}{\beta^2 a}\sin(\beta \mathrm{a}) \\ -\beta\sin(\beta a) & -\sin(\beta \mathrm{a}) \end{bmatrix} \\ & + (\Delta\beta_i a)^2\begin{bmatrix} -\frac{1}{2}\cos(\beta a) & \frac{1}{\beta^3 a^2}\sin(\beta \mathrm{a}) - \frac{1}{2\beta}\sin(\beta \mathrm{a}) - \frac{1}{\beta^2 a}\cos(\beta a) \\ \frac{1}{2}\beta\sin(\beta a) - \frac{1}{a}\cos(\beta a) & -\frac{1}{2}\cos(\beta a) \end{bmatrix}. \end{aligned} \tag{3.76}$$

For the sake of simplicity, let us represent the three matrices in Eq. (3.76) by A, B, and C, respectively, so Eq. (3.76) can be recast into

$$M_i = A + (\Delta\beta_i a)B + (\Delta\beta_i a)^2 C. \tag{3.77}$$

The matrix of a whole perturbation period takes the form

$$M(\Lambda) = \underbrace{M_l M_{l-1} \ldots M_2 M_1}_{l}. \tag{3.78}$$

And as shown in the last section, in order to calculate the coupling coefficient, the rank of the transfer matrix should be obtained in advance. Two general laws on the matrix rank are listed below:

$$\mathrm{Tr}(M+N) = \mathrm{Tr}(M) + \mathrm{Tr}(N), \tag{3.79}$$

$$\mathrm{Tr}(MN) = \mathrm{Tr}(NM). \tag{3.80}$$

Using the two relations above, and ignoring all the terms whose order is higher than $(\Delta\beta_i a)^3$, we can derive the rank corresponding to the transfer matrix of the perturbation period, which is written as

$$\begin{aligned}\mathrm{Tr}(M(\Lambda)) &= \mathrm{Tr}(A^l) + \mathrm{Tr}(BA^{l-1})\sum_i(\Delta\beta_i a)\\&+\mathrm{Tr}(CA^{l-1})\sum_i(\Delta\beta_i a)^2 + \frac{1}{2}\sum_j\sum_i(\Delta\beta_i a)(\Delta\beta_i a)\mathrm{Tr}(BA^{j-i-1}BA^{l-j+i-1})\\&(j=1,2,\ldots,l;\ \ i=1,2,\ldots,l),\end{aligned} \tag{3.81}$$

where $\mathrm{Tr}(A^l)$ is the rank of the product of all the matrix A, which is the first term of the matrix M_i, and $\mathrm{Tr}(BA^{l-1})\sum_i(\Delta\beta_i a)$ is the rank of the product of $(l-1)$ matrix A and one matrix B; similarly, $\mathrm{Tr}(CA^{l-1})\sum_i(\Delta\beta_i a)^2$ corresponds to the product of $(l-1)$ matrix A and one matrix C. Let us consider the fourth term in Eq. (3.81) in detail, which is determined by the product of $(l-2)$ matrix A and two matrix B. If the two matrixes B are referred as index of i and j, which are arbitrarily chosen, then the array of all matrixes in the multiplication can be written as

$$\underbrace{AA\ldots A}_{(i-1)}\underbrace{B}_{i}\underbrace{AA\ldots A}_{(j-i-1)}\underbrace{B}_{j}\underbrace{AA\ldots A}_{(l-j)}$$

If we swap i and j, the array will be the same, so in order to eliminate double counting, a factor of $1/2$ must be introduced.

Let us calculate Eq. (3.81) term by term, and according to Eqs. (3.23), (3.24), and (3.25), we can write

$$\mathrm{Tr}(A^l) = 2\cos\beta\Lambda. \tag{3.82}$$

Meanwhile, when $l \to \infty$, there is $a \to 0$, so in view of Eqs. (3.72) and (3.75), we obtain

$$\mathrm{Tr}(BA^{l-1})\sum_i(\Delta\beta_i a) = \frac{\kappa^2}{\beta w_{\mathrm{eff}}}\mathrm{Tr}(BA^{l-1})\int_0^\Lambda f(z)\mathrm{d}z = 0. \tag{3.83}$$

For the third term on the right side of Eq. (3.81), when $l \to \infty$, it can be written in the form of below integration

$$\mathrm{Tr}(CA^{l-1})\sum_i(\Delta\beta_i a)^2 = \left(\frac{\kappa^2}{\beta w_{\mathrm{eff}}}\right)^2\mathrm{Tr}(CA^{l-1})\cdot a\int_0^\Lambda f^2(z)dz, \tag{3.84}$$

which also approaches zero as $a \to 0$.

If we consider the foruth term on the right side of Eq. (3.81), by matrix operations, it is not difficult to find

$$\begin{aligned}\mathrm{Tr}\left(BA^{j-i-1}BA^{l-j+i-1}\right) &= -2\cos\beta\Lambda + 2\left(\frac{\sin\beta a}{\beta a}\right)^2\cos[l-2(j-i)]\beta a \\ &= 2-2\cos[2(j-i)]\beta a.\end{aligned} \tag{3.85}$$

In the above calculation, we already make use of the approximation $\beta\Lambda \approx \pi$ and the below limit

$$\lim_{a\to 0}\frac{\sin\beta a}{\beta a} = 1.$$

Consequently, there is

$$\begin{aligned}&\frac{1}{2}\sum_j\sum_i(\Delta\beta_i a)(\Delta\beta_i a)\mathrm{Tr}\left(BA^{j-i-1}BA^{l-j+i-1}\right. \\ &=\sum_j\sum_i(\Delta\beta_i a)(\Delta\beta_i a)\{1-\cos[2(j-i)\beta a]\}.\end{aligned} \tag{3.86}$$

In order to turn the above summations into integration, let us take the limit $l\to\infty$, $a\to 0$. Set $ia = z_1$, $ja = z_2$, and $\Delta i = \Delta j = 1$, and then, Eq. (3.86) can be transformed into

$$\begin{aligned}&\left(\frac{\kappa^2}{\beta w_{\mathrm{eff}}}\right)^2\int_0^\Lambda\int_0^\Lambda f(z_1)f(z_2)\{1-\cos[2\beta(z_2-z_1)]\}\mathrm{d}z_1\mathrm{d}z_2 \\ &= -\left(\frac{\kappa^2}{\beta w_{\mathrm{eff}}}\right)^2\int_0^\Lambda\int_0^\Lambda f(z_1)f(z_2)\cos[2\beta(z_2-z_1)]\mathrm{d}z_1\mathrm{d}z_2 \\ &= -\left(\frac{\kappa^2}{\beta w_{\mathrm{eff}}}\right)^2\int_0^\Lambda\int_0^\Lambda f(z_1)f(z_2)\frac{\mathrm{e}^{i2\beta(z_2-z_1)}-\mathrm{e}^{-i2\beta(z_2-z_1)}}{2}\mathrm{d}z_1\mathrm{d}z_2 \\ &= -\left(\frac{\kappa^2}{\beta w_{\mathrm{eff}}}\right)^2\int_0^\Lambda\int_0^\Lambda f(z_1)f(z_2)\mathrm{e}^{i2\beta(z_2-z_1)}\mathrm{d}z_1\mathrm{d}z_2 \\ &= -\left(\frac{\kappa^2}{\beta w_{\mathrm{eff}}}\right)^2\left|\int_0^\Lambda f(z)\mathrm{e}^{i2\beta z}\mathrm{d}z\right|^2,\end{aligned} \tag{3.87}$$

which yields

$$\begin{aligned}\chi &= \frac{1}{2}\mathrm{Tr}[M(\Lambda)]^2 \\ &= \cos\beta\Lambda - \frac{1}{2}\left|\left(\frac{\kappa^2}{\beta w_{\mathrm{eff}}}\right)\int_0^\Lambda f(z)\mathrm{e}^{i2\beta z}\mathrm{d}z\right|^2.\end{aligned} \tag{3.88}$$

Using the Eqs. (3.28) and (3.39), we finally arrive at the distributed feedback coefficient for corrugated periodic waveguide of arbitrary shape

$$\kappa_c = \frac{\kappa^2}{\beta\Lambda w_{\mathrm{eff}}}\left|\int_0^\Lambda f(z)\mathrm{e}^{i2\beta z}\mathrm{d}z\right|. \tag{3.89}$$

Replacing $\beta\Lambda$ with π, the above formula becomes

$$\kappa_c = \frac{\kappa^2}{\pi w_{\mathrm{eff}}}\left|\int_0^\Lambda f(z)\mathrm{e}^{i2\beta z}\mathrm{d}z\right|. \tag{3.90}$$

The phase factor in the above integration shows that the derived distributed feedback coefficient can be applied to describe the energy transfer between the forward and backward traveling waves since the latter wave experiences a phase shift twice that of the former wave. So the physical insight of Eq. (3.90) is rather clear. Besides, when compared with the coupled wave analysis, three apparent differences can be observed:

1. Equation (3.90) is the integration along the direction of the perturbation period instead of its perpendicular direction.
2. The integration is independent of the transverse eigenmode field in the unperturbed waveguide.
3. The coupling coefficient is related to the corrugated shape function of the waveguide, and it is not related to the Fourier coefficients of the shape function.

Based on the above characteristics, the expression of the distributed feedback coefficient derived from the transfer matrix method is simple, and the calculation for a specific corrugated shape function is not difficult. Analytical expression can be easily derived, and no numerical simulation is required.

3.2.2 Typical Corrugated Periodic Waveguide

This section will discuss four kinds of typical corrugated periodic waveguide, whose distributed feedback coefficients are calculated via the integration. The results are satisfactory when compared with numerical simulations.

1. Rectangular corrugated periodic waveguide (Fig. 3.6)

$$f(z) = \begin{cases} g_2, & 0 \le z < t, \\ -g_1, & t \le z < \Lambda. \end{cases} \tag{3.91}$$

Inserting Eq. (3.91) into (3.90), one can obtain

$$\kappa_c = \frac{\kappa^2 g}{\pi \beta w_{\text{eff}}} \sin(\beta t). \tag{3.92}$$

when $t = \Lambda/2$, there is

$$\kappa_c = \frac{\kappa^2 g}{\pi \beta w_{\text{eff}}}. \tag{3.93}$$

In view of $\beta\Lambda = \pi$, it is clear that Eq. (3.93) is the same result with that of Eq. (3.41).

2. Sinusoidal corrugated periodic waveguide

The sinusoidal corrugated periodic waveguide is shown in Fig. 3.7, whose corrugated function is given by

$$f(z) = \frac{g}{2} \sin\left(\frac{2\pi}{\Lambda} z\right), \tag{3.94}$$

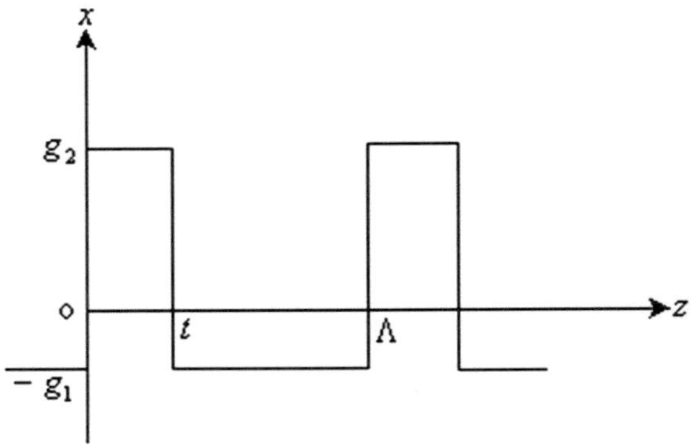

Fig. 3.6 Rectangular corrugated periodic waveguide

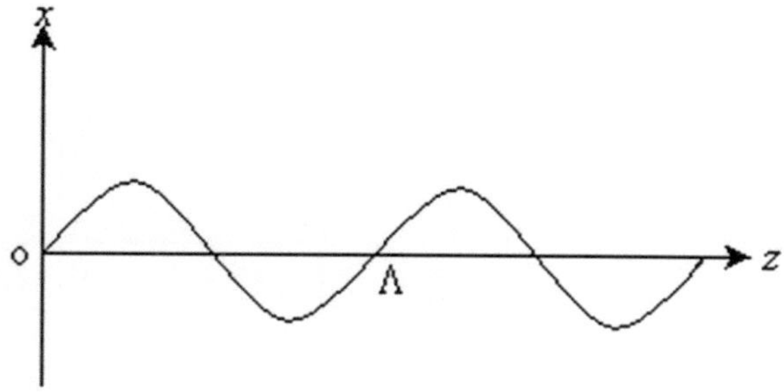

Fig. 3.7 Sinusoidal corrugated periodic waveguide

Inserting Eq. (3.94) into Eq. (3.90), it yields

$$\kappa_c = \frac{\kappa^2 g}{4\beta w_{\text{eff}}}. \tag{3.95}$$

3. Symmetrical triangle corrugated waveguide
   The symmetrical triangle corrugated waveguide is plotted in Fig. 3.8, whose corrugated function takes the form of

$$f(z) = \begin{cases} \frac{2g}{\Lambda} z, & 0 \le z < \frac{\Lambda}{4}, \\ g - \frac{2g}{\Lambda} z, & \frac{\Lambda}{4} \le z < \frac{3\Lambda}{4}, \\ \frac{2g}{\Lambda} z - 2g, & \frac{3\Lambda}{4} \le z < \Lambda. \end{cases} \tag{3.96}$$

   The following result can be obtained via substituting Eq. (3.96) into (3.90)

$$\kappa_c = \frac{2\kappa^2 g}{\pi^2 \beta w_{\text{eff}}}. \tag{3.97}$$

4. Zigzag corrugated periodic waveguide

The zigzag corrugated periodic waveguide is plotted in Fig. 3.9, whose corrugated shape can be described by

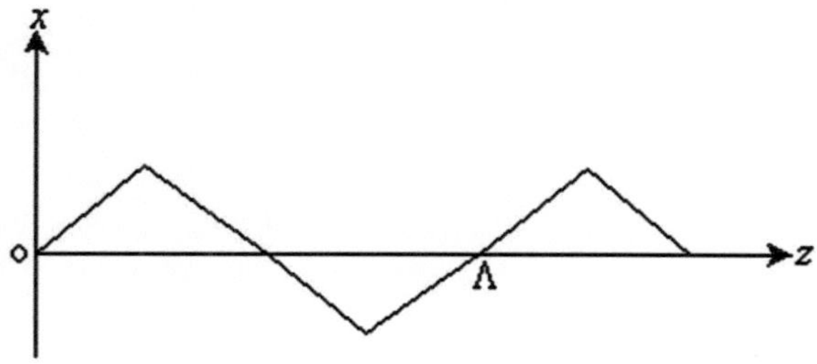

Fig. 3.8 Symmetrical triangle corrugated waveguide

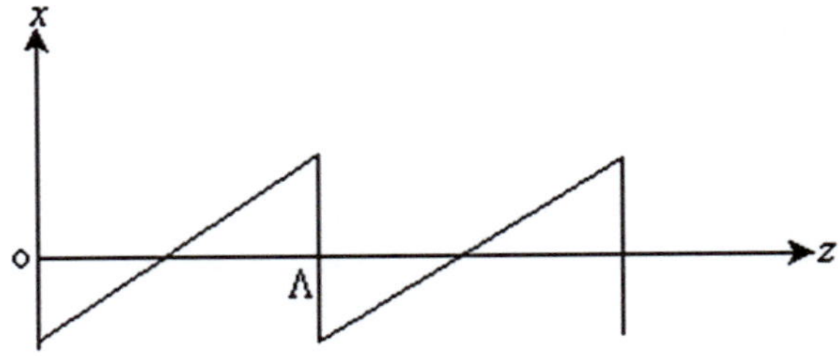

Fig. 3.9 Zigzag corrugated periodic waveguide

$$f(z) = \frac{g}{\Lambda}z - \frac{g}{2}. \tag{3.98}$$

Substituting Eq. (3.98) into (3.90), it is clear that

$$\kappa_c = \frac{\kappa^2 g}{2\pi\beta w_{\text{eff}}}. \tag{3.99}$$

As can be seen from the above discussion, the difference between the distributed feedback coefficients of the four kinds of typical corrugated periodic waveguide is just a constant factor. All the results derived in this section are less complicated than those derived via the coupled wave analysis. And this calculation process is much simpler, and also, the clear physical insight is provided.

Figure 3.10 plots the distributed feedback coefficient of the TE_0 mode of the four structures as a function of the corrugated height variation. The parameters used in the plotting are provided below:

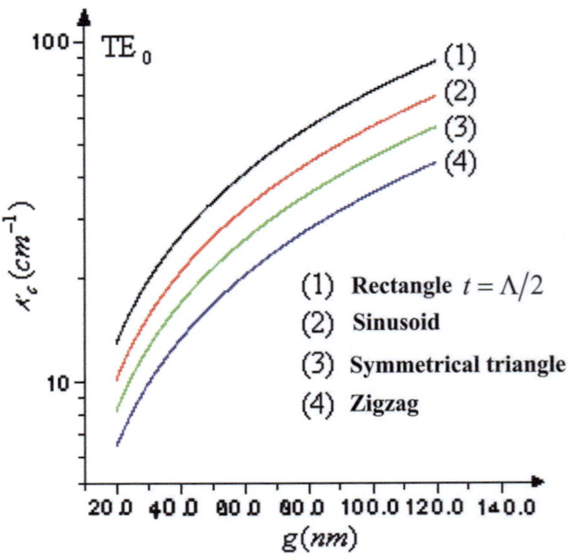

Fig. 3.10 The coupling coefficient curve as a function of the corrugated height

$$n_2 = n_3 = 3.4,$$
$$n_1 = 3.6,$$
$$\lambda = 0.85\ \mu\text{m},$$
$$w = 1\ \mu\text{m}.$$

It can be seen from the figure, the rectangular corrugated waveguide has the highest coupling coefficient, while the zigzag corrugated waveguide has the lowest coefficient.

3.3 Step-Index Multiple Quantum Well (MQW) Optical Waveguide

The dispersion of the step-index multiple quantum well (MQW) optical waveguide is analyzed by the transfer matrix method, while the effective index is also derived as an analytical expression.

3.3.1 Effective Permittivity of the Infinite Periodic Multilayers [4]

Let us consider the periodic multilayer structure, which is composed of two media with permittivity ε_1 and ε_2, respectively, as illustrated in Fig. 3.11. If the film thickness of the two media are h_1 and h_2, then the period is $\Lambda = h_1 + h_2$. For the semiconductor MQW or superlattice, the period is composed of several atom layers. From the view of symmetry, the permittivity of these structures has the same form of the uniaxial crystal, i.e., $\varepsilon_y = \varepsilon_z = \varepsilon_\perp$, $\varepsilon_x = \varepsilon_{//}$. As a result, the y (or z)-component of the averaged electric displacement vector in a period is

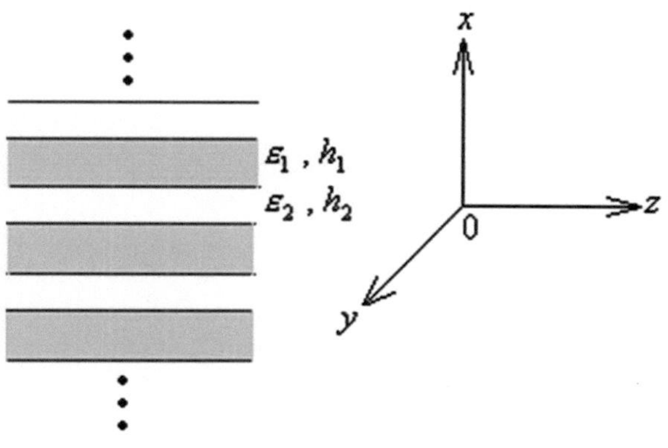

Fig. 3.11 Infinitely long periodic multilayers

$$\overline{D_y} = \frac{1}{h_1 + h_2}\left(\varepsilon_1 E_y^{(1)} h_1 + \varepsilon_2 E_y^{(2)} h_2\right), \tag{3.100}$$

where $E_y^{(1)}$ and $E_y^{(2)}$ are the y-component of the electric field in the medium 1 and 2, respectively. Since both $E_y^{(1)}$ and $E_y^{(2)}$ are continuous on the interface, and $h_{1,2} \ll \lambda$ (λ represents the optical wavelength), the electric field variation in the thin film can be ignored and one can yield

$$E_y^{(1)} = E_y^{(2)} = \overline{E_y}, \tag{3.101}$$

where $\overline{E_y}$ is the average of the electric field in a period, and $\overline{D_y}$ is determined by

$$\overline{D_y} = \varepsilon_\perp \overline{E_y},$$

where

$$\varepsilon_\perp = \frac{1}{h_1 + h_2}(\varepsilon_1 h_1 + \varepsilon_2 h_2). \tag{3.102}$$

Similarly, the x-component E_x of the electric field in a period has an average value as

$$\overline{E_x} = \frac{1}{h_1 + h_2}\left(\frac{D_x^{(1)}}{\varepsilon_1} h_1 + \frac{D_x^{(2)}}{\varepsilon_2} h_2\right). \tag{3.103}$$

In view of the continuity of D_x at the interface, and ignoring its variation in the thin film, we have

$$D_x^{(1)} = D_x^{(2)} = \overline{D_x}. \tag{3.104}$$

Combining the above equation with

$$\overline{E_x} = \varepsilon_{//}^{-1} \overline{D_x},$$

it is clear that

$$\frac{1}{\varepsilon_{//}} = \frac{1}{h_1 + h_2}\left(\frac{h_1}{\varepsilon_1} + \frac{h_2}{\varepsilon_2}\right). \tag{3.105}$$

Equations (3.102) and (3.105) are the effective index expressions for TE and TM polarizations of the infinitely long periodic multilayer structures, respectively. It is easy to see that this structure is similar to the uniaxis crystal and demonstrates the effect of birefringence. $\varepsilon_\perp$ corresponds to the permittivity of the ordinary light, while $\varepsilon_{//}$ corresponds to the extraordinary light.

3.3.2 *Effective Index of the MQW Waveguide*

MQW optical waveguide and the infinitely long periodic multilayer structure are different. But when dealing with MQW optical waveguide, its effective index is usually approximated by Eqs. (3.102) and (3.105) [5]. This approximation is not reasonable, so Ohke [6, 7] and Skinner [8] try to solve this problem by a different approach, but their methods can only be applied to the TE mode in the MQW optical waveguide, while the TM mode remains unsolved. Combining the transfer matrix and Floquet theory, we derived the effective index for both polarizations of the MQW optical waveguide in the thin-film approximation and explained the birefringence effect in the MQW optical waveguide [9].

Let us consider the refractive index distribution of the MQW optical waveguide as plotted in Fig. 3.12, in which the square of the refractive index is plotted for convenience. Assuming the thickness of the film with refractive index n_1 is h_1, while h_2 is the film thickness with refractive index n_2, so the total period is given by $\Lambda = h_1 + h_2$, and the thickness of the whole quantum well region is

$$w = N\Lambda + h_1, \tag{3.106}$$

where N is the cell number of the quantum well region. n_0 and n_3 are the refractive index of the cover layer and the substrate, respectively. Without loss of generality, we assume that

$$n_1 > n_3 > n_2 > n_0. \tag{3.107}$$

Based on the electromagnetic theory of the dielectric slab waveguide, the electric field decays exponentially in the cover layer and the substrate, so we assume their exponential decay coefficients are q and p, respectively. In the h_1 thick film, the electric field is the superposition of two oscillating waves traveling in opposite directions, while in the h_2 thick film, the field is represented by the superposition of

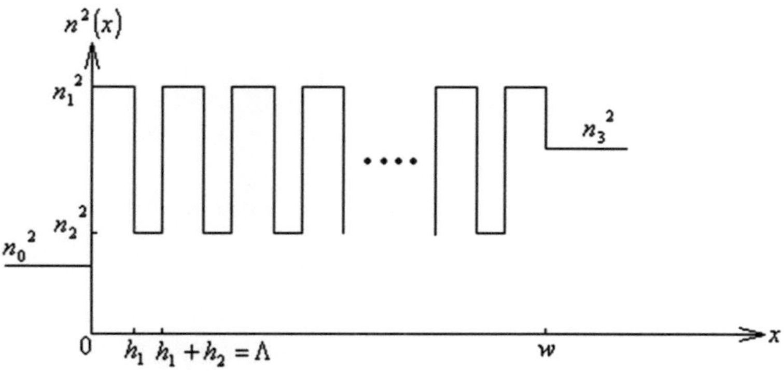

Fig. 3.12 Step-index MQW optical waveguide

two exponential decay fields. Defining the characteristic parameter of the oscillating and exponential decay field to be κ and α, it is easy to prove that the four characteristic parameters are given by

$$\left.\begin{array}{l} q = \left(\beta^2 - k_0^2 n_0^2\right)^{1/2} \\ p = \left(\beta^2 - k_0^2 n_3^2\right)^{1/2} \\ \kappa = \left(k_0^2 n_1^2 - \beta^2\right)^{1/2} \\ \alpha = \left(\beta^2 - k_0^2 n_2^2\right)^{1/2} \end{array}\right\}. \tag{3.108}$$

The transfer matrix representing the h_1 thick film is

$$M(h_1) = \begin{bmatrix} \cos(\kappa h_1) & -\frac{f_1}{\kappa}\sin(\kappa h_1) \\ \frac{\kappa}{f_1}\sin(\kappa h_1) & \cos(\kappa h_1) \end{bmatrix}. \tag{3.109}$$

Since the h_2 thick film supports two exponential decay fields, its transfer matrix takes the form

$$M(h_2) = \begin{bmatrix} \cosh(\alpha h_2) & -\frac{f_2}{\alpha}\sinh(\alpha h_2) \\ -\frac{\alpha}{f_2}\sinh(\alpha h_2) & \cosh(\alpha h_2) \end{bmatrix}, \tag{3.110}$$

where

$$f_j = \left| \begin{array}{ll} 1, & \text{TE mode}, \\ n_j^2, & \text{TM mode}, \quad (\mathrm{j} = 0, 1, 2, 3). \end{array} \right. \tag{3.111}$$

Combining the above two matrixes, the matrix corresponds to a whole period Λ is

$$M(\Lambda) = M(h_1) \cdot M(h_2), \tag{3.112}$$

which can also be expressed as

$$M(\Lambda) = \begin{bmatrix} m_{11} & m_{12} \\ m_{21} & m_{22} \end{bmatrix}, \tag{3.113}$$

where

$$m_{11} = \cos(\kappa h_1)\cosh(\alpha h_2) + \frac{\alpha f_1}{\kappa f_2}\sin(\kappa h_1)\sinh(\alpha h_2), \tag{3.114}$$

$$m_{12} = -\frac{f_1}{\kappa}\sin(\kappa h_1)\cosh(\alpha h_2) - \frac{f_2}{\alpha}\cos(\kappa h_1)\sinh(\alpha h_2), \tag{3.115}$$

$$m_{21} = \frac{\kappa}{f_1}\sin(\kappa h_1)\cosh(\alpha h_2) - \frac{\alpha}{f_2}\cos(\kappa h_1)\sinh(\alpha h_2), \tag{3.116}$$

$$m_{22} = \cos(\kappa h_1)\cosh(\alpha h_2) - \frac{\kappa f_2}{\alpha f_1}\sin(\kappa h_1)\sinh(\alpha h_2). \tag{3.117}$$

Based on the Floquet theory, the propagation constant K in the periodic structure obeys

$$\cos(K\Lambda) = \frac{1}{2}\mathrm{Tr}[M(\Lambda)]. \tag{3.118}$$

Based on Eqs. (3.114) and (3.117), it can be derived that

$$\begin{aligned}\cos(K\Lambda) &= \frac{1}{2}(m_{11} + m_{22}) \\ &= \cos(\kappa h_1)\cosh(\alpha h_2) + \frac{1}{2}\left(\frac{\alpha f_1}{\kappa f_2} - \frac{\kappa f_2}{\alpha f_1}\right)\sin(\kappa h_1)\sinh(\alpha h_2),\end{aligned} \tag{3.119}$$

and the wave number of the Bloch wave is

$$K = \left(k_0^2 n_e^2 - \beta^2\right)^{1/2}, \tag{3.120}$$

where the n_e is clearly the effective index of the core layer of the MQW optical waveguide.

In the typical MQW waveguide, the core layer usually contains fifty to several hundred periods, whose length Λ is around 10 nm. Therefore, the inequality $h_{1,2} \ll \lambda$ (λ is the optical wavelength) holds, and the characteristic parameters K, κ, and α are of the scale of the reciprocal of wavelength. Expanding the trigonometric and hyperbolic functions in Eq. (3.119) and taking the second-order approximation, it follows

$$1 - \frac{1}{2}(K\Lambda)^2 = \left[1 - \frac{1}{2}(\kappa h_1)^2\right]\left[1 + \frac{1}{2}(\alpha h_2)^2\right] + \frac{1}{2}h_1 h_2 \frac{\alpha^2 f_1^2 - \kappa^2 f_2^2}{f_1 f_2}. \tag{3.121}$$

Using Eqs. (3.108) and (3.120), the effective index of the core layer of the MQW waveguide can be derived from Eq. (3.121) as

$$n_e^2 = \frac{n_1^2 h_1 + n_2^2 h_2}{h_1 + h_2} - \frac{h_1 h_2}{(h_1 + h_2)^2}\cdot\frac{(f_1 - f_2)^2}{f_1 f_2}(\beta/k_0)^2. \tag{3.122}$$

Based on the above discussion, the core layer of the MQW optical waveguide can be treated as a single film with refractive index n_e, and the MQW optical waveguide can be treated as a simple three-layer slab waveguide. If one defines

$$f_e = \begin{cases} 1, & \text{TE mode}, \\ n_e^2, & \text{TM mode}, \end{cases} \tag{3.123}$$

then the transfer matrix of the core layer of the MQW optical waveguide is

$$M(w) = \begin{bmatrix} \cos(Kw) & -\frac{f_e}{K}\sin(Kw) \\ \frac{K}{f_e}\sin(Kw) & \cos(Kw) \end{bmatrix}. \tag{3.124}$$

With the help of Eqs. (2.42) and (2.43), the dispersion relation of the MQW optical waveguide can be cast into

$$\begin{bmatrix} -\frac{p}{f_3} & 1 \end{bmatrix} M(w) \begin{bmatrix} 1 \\ -\frac{q}{f_0} \end{bmatrix} = 0. \tag{3.125}$$

By simplification, one has

$$\left(\frac{K^2}{f_e^2} - \frac{pq}{f_0 f_3}\right)\tan(Kw) = \frac{K}{f_e}\left(\frac{p}{f_3} + \frac{q}{f_0}\right). \tag{3.126}$$

Combining with Eqs. (3.122) and (3.120), the transcendental Eq. (3.126) can completely specify the propagation and dispersion characteristics of the MQW optical waveguide under two polarizations.

Let us discuss the physics behind Eq. (3.122) in more detail.

1. For the TE mode, according to Eq. (3.122), the effective index of the TE mode in the core layer of the MQW optical waveguide is

$$\left(n_e^{\mathrm{TE}}\right)^2 = \frac{n_1^2 h_1 + n_2^2 h_2}{h_1 + h_2}, \tag{3.127}$$

   which is identical with the effective permittivity Eq. (3.102) of the infinitely long periodic multilayer structure. This is reasonable; since the boundary condition exerts no influence for TE mode, the effective index of the TE mode is the same for the MQW optical waveguide and the infinitely long periodic multilayer structure.
2. For TM mode, the effective index of the TM mode in the core layer of the MQW optical waveguide can be obtained from Eq. (3.122)

$$\left(n_e^{\mathrm{TM}}\right)^2 = \frac{n_1^2 h_1 + n_2^2 h_2}{h_1 + h_2} - \frac{h_1 h_2}{(h_1 + h_2)^2} \cdot \frac{\left(n_1^2 - n_2^2\right)^2}{n_1^2 n_2^2} (\beta/k_0)^2. \tag{3.128}$$

   It is clear that the effective index of the TM mode is no longer constant, but varies as a function of the characteristic parameter of the optical waveguide, i.e., the refractive index (β/k_0). There was some misunderstanding on this point [6], and some attempt has been made to modify Eq. (3.128). In fact, TM mode has an electric field component E_x, which is perpendicular to the waveguide interface, so it is natural that different boundary condition will affect this component. It is obvious that the effective index of the TM mode should be connected to the parameters of the waveguide.

3. When $w \to \infty$, the MQW optical waveguide turns into the infinitely long periodic multilayer structure. If the thickness of the core layer in a three-layer slab waveguide increases, the effective refractive index of the waveguide increases as well. Consequently, if the core layer thickness of the MQW optical waveguide increases, its refractive index (β/k_0) also increases. And when w approaches infinite, we have

$$\beta/k_0 \to \left(n_e^{\mathrm{TM}}\right)_\infty, \tag{3.129}$$

where the subscript "∞" denotes the limit when the core layer thickness approaches infinite. Meanwhile, $\left(n_e^{\mathrm{TM}}\right)_\infty$ becomes the effective index of the TM mode of the infinitely long periodic multilayer structure. Inserting Eq. (3.129) into (3.128), and notice that the n_e^{TM} in the left-hand side of Eq. (3.128) should be replaced by $\left(n_e^{\mathrm{TM}}\right)_\infty$, one can write

$$\left(n_e^{\mathrm{TM}}\right)_\infty^2 = \frac{h_1 + h_2}{\frac{h_1}{n_1^2} + \frac{h_2}{n_2^2}}, \tag{3.130}$$

which is exactly the same as Eq. (3.105) and confirms that the effective index given by Eq. (3.128) is correct.

4. If the two media in the core layer have the same refractive index, i.e., $n_1 = n_2$, the MQW optical waveguide reduces to a simple three-layer slab waveguide, and Eq. (3.128) becomes

$$n_e^{\mathrm{TM}} = n_e^{\mathrm{TE}}. \tag{3.131}$$

This is due to the core layer effective index is always constant for the simple three-layer slab waveguide and independent of the light polarization.

5. If $n_1 \neq n_2$, it is easy to see from Eq. (3.128) that

$$n_e^{\mathrm{TM}} \neq n_e^{\mathrm{TE}}. \tag{3.132}$$

This is the feature that MQW optical waveguide is different from the three-layer slab waveguide. And this inequality is the physical reason for the birefringence effect of the MQW optical waveguide.

6. If the refractive index n_3 of the substrate of the MQW optical waveguide fulfills the following inequality

$$n_e^{\mathrm{TM}} < n_3 < n_e^{\mathrm{TE}}, \tag{3.133}$$

then the MQW optical waveguide can only support TE mode. So Eq. (3.133) provides a new strategy to design a novel polarizer based on optical waveguide.

3.4 MQW Optical Waveguide with Arbitrary Refractive Index Distribution

The above section has analyzed the simple MQW optical waveguide with a step-index distribution. But due to the various factors in the manufacturing process, the refractive index distribution of the MQW optical waveguide is usually continuous. So in this section, we will deal with arbitrary refractive index distribution. Two methods, including the effective index approximation and the non-effective index method, are developed.

3.4.1 Effective Index Method [10]

Let us consider the MQW optical waveguide with an arbitrary refractive index profile as plotted in Fig. 3.13, and assume the light travels along the z-axis. The period of the quantum well (barrier) is set to be Λ, and the length of the MQW region is w, which includes N periods. The refractive index of the cover layer and substrate is n_0 and n_3, respectively, while the refractive index of the core layer ranges from n_2 to n_1. The refractive index profile of a period is given by the function $n(x)$, which obeys the periodic condition

$$n(x+\Lambda) = n(x). \tag{3.134}$$

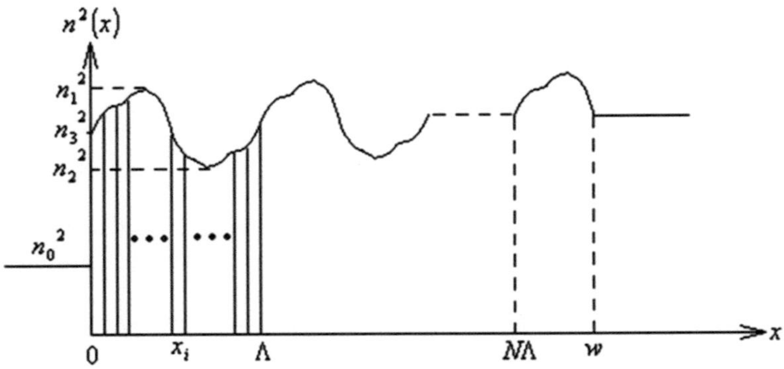

Fig. 3.13 MQW optical waveguide with arbitrary refractive index profile

Without loss of generality, let us assume

$$n_1 > n_3 > n_2 > n_0. \tag{3.135}$$

In order to apply the transfer matrix, we divide a period Λ into l sections with an equal width. If the width of a small section is h, then there is $x_l = lh = \Lambda$. Generally, the transfer matrix for the ith section can be written as

$$M_i = \begin{bmatrix} \cos(\kappa_i h) & -\frac{f_i}{\kappa}\sin(\kappa_i h) \\ \frac{\kappa_i}{f_i}\sin(\kappa_i h) & \cos(\kappa_i h) \end{bmatrix}, \quad (i = 1, 2, \ldots l), \tag{3.136}$$

where

$$f_i = \begin{cases} 1, & \text{TE mode}, \\ n^2(x_i), & \text{TM mode}, \end{cases} \tag{3.137}$$

$$\kappa_i = \left[k_0^2 n^2(x_i) - \beta^2\right]^{1/2}. \tag{3.138}$$

Clearly, κ_i is a function of x, and it can be real or complex. When $k_0 n(x_i) > \beta$, the electric field in the ith section is oscillating, and the matrix given by Eq. (3.136) is real. When $\beta > k_0 n(x_i)$, κ should be replaced by $i\alpha$ due to the exponential decay field. If we applied the following formula

$$\begin{aligned} \cos(i\alpha h) &= \cosh(\alpha h), \\ \sin(i\alpha h) &= i\,\sinh(\alpha h), \end{aligned}$$

the corresponding matrix can still be represented by Eq. (3.136).

In case of $N \gg 1$, Λ is usually of the scale of 100 Å, and when $l \to \infty$, there is $h \to 0$, so the following condition holds

$$\kappa_i h \ll 1. \tag{3.139}$$

And the thin-film approximation can be applied, expanding the sine and cosine functions in the matrix M_i into series, and ignoring all the terms higher than the second order, we can rewrite Eq. (3.136) as

$$\begin{aligned} M_i &= \begin{bmatrix} 1 - \frac{1}{2}(\kappa_i h)^2 & -f_i h \\ \frac{\kappa_i^2}{f_i} h & 1 - \frac{1}{2}(\kappa_i h)^2 \end{bmatrix} \\ &= \begin{bmatrix} 1 & 0 \\ 0 & 1 \end{bmatrix} - f_i h \begin{bmatrix} 0 & 1 \\ 0 & 0 \end{bmatrix} + \frac{\kappa_i^2}{f_i} h \begin{bmatrix} 0 & 0 \\ 1 & 0 \end{bmatrix} - \frac{1}{2}(\kappa_i h)^2 \begin{bmatrix} 1 & 0 \\ 0 & 1 \end{bmatrix}. \end{aligned} \tag{3.140}$$

Clearly, the first term on the right-hand side is the dominant term, while the second and third terms are the first-order infinitesimal, and the fourth term is the second-order infinitesimal. Denote that

$$M_i = E - (f_i h)A + \left(\frac{\kappa_i^2}{f_i}h\right)B - \frac{1}{2}(\kappa_i h)^2 E, \tag{3.141}$$

where E represents the unit matrix, and A and B are given by

$$E = \begin{bmatrix} 1 & 0 \\ 0 & 1 \end{bmatrix}, \quad A = \begin{bmatrix} 0 & 1 \\ 0 & 0 \end{bmatrix}, \quad B = \begin{bmatrix} 0 & 0 \\ 1 & 0 \end{bmatrix}.$$

The transfer matrix for a period Λ is the product of l matrix, i.e., $M_i (i = 1, 2, \cdots, l)$, so we have

$$M(\Lambda) = M_1 M_2 \ldots M_{l-1} M_l. \tag{3.142}$$

The trace of the sum or product of two matrixes M and N should obey the following rules:

$$\begin{aligned} \mathrm{Tr}(M+N) &= \mathrm{Tr}(M) + \mathrm{Tr}(N), \\ \mathrm{Tr}(MN) &= \mathrm{Tr}(NM). \end{aligned}$$

So using the feature of the matrix trace and inserting Eq. (3.141) into (3.142), the trace of the transfer matrix corresponds to a period Λ can be cast into:

$$\begin{aligned} \mathrm{Tr}[M(\Lambda)] = {} & \mathrm{Tr}(E^l) - \sum_i f_i h \mathrm{Tr}(E^{l-1}A) + \sum_i \left(\frac{\kappa_i^2}{f_i}h\right)\mathrm{Tr}(E^{l-1}B) \\ & - \frac{1}{2}\sum_i (\kappa_i h)^2 \mathrm{Tr}(E^l) + \frac{1}{2}\sum_i \sum_{\substack{j \\ (i \neq j)}} (f_i h)(f_j h)\mathrm{Tr}\left(AE^{j-i-1}AE^{l-j+i-1}\right) \\ & + \frac{1}{2}\sum_i \sum_{\substack{j \\ (i \neq j)}} \left(\frac{\kappa_i^2}{f_i}h\right)\left(\frac{\kappa_j^2}{f_j}h\right)\mathrm{Tr}\left(BE^{j-i-1}BE^{l-j+i-1}\right) \\ & - \sum_i \sum_{\substack{j \\ (i \neq j)}} (f_i h)\left(\frac{\kappa_j^2}{f_j}h\right)\mathrm{Tr}\left(AE^{j-i-1}BE^{l-j+i-1}\right). \end{aligned} \tag{3.143}$$

Below, we calculate each term on the right-hand side of Eq. (3.143). Based on the characteristics of the unit matrix E and two special matrixes A and B, it is easy to see

$$\begin{aligned}
&\mathrm{Tr}\left(E^{l}\right) = \mathrm{Tr}(E) = 2,\\
&\mathrm{Tr}\left(E^{l-1}A\right) = \mathrm{Tr}(A) = 0,\\
&\mathrm{Tr}\left(E^{l-1}B\right) = \mathrm{Tr}(B) = 0,\\
&\mathrm{Tr}\left(AE^{j-i-1}AE^{l-j+i-1}\right) = \mathrm{Tr}\left(A^{2}\right) = 0,\\
&\mathrm{Tr}\left(BE^{j-i-1}BE^{l-j+i-1}\right) = \mathrm{Tr}\left(B^{2}\right) = 0,\\
&\mathrm{Tr}\left(AE^{j-i-1}AE^{l-j+i-1}\right) = \mathrm{Tr}(AB) = 1.
\end{aligned}$$

Using the above conclusion, Eq. (3.143) turns into

$$\mathrm{Tr}[M(\Lambda)] = 2 - \sum_{i} (\kappa_i h)^2 - \sum_{i} \sum_{\substack{j \\ (i \neq j)}} (f_i h) \left(\frac{\kappa_j^2}{f_j} h \right). \tag{3.144}$$

The second and third terms on the right-hand side of Eq. (3.144) are the sum of the diagonal elements and the rest elements, respectively. In order to be more specific, considering the case of $l = 3$, the matrix takes the form:

$$\begin{bmatrix}
f_1 h \cdot \frac{\kappa_1^2}{f_1} h & f_1 h \cdot \frac{\kappa_2^2}{f_2} h & f_1 h \cdot \frac{\kappa_3^2}{f_3} h \\
f_2 h \cdot \frac{\kappa_1^2}{f_1} h & f_2 h \cdot \frac{\kappa_2^2}{f_2} h & f_2 h \cdot \frac{\kappa_3^2}{f_3} h \\
f_3 h \cdot \frac{\kappa_1^2}{f_1} h & f_3 h \cdot \frac{\kappa_2^2}{f_2} h & f_3 h \cdot \frac{\kappa_3^2}{f_3} h
\end{bmatrix}. \tag{3.145}$$

It is clear that when $l \to \infty$, there is $h \to 0$, so the sum of the matrix elements becomes a double integral. Using Eqs. (3.137) and (3.136), we can derive the below results.

For the TE mode

$$\begin{aligned}
&\sum_{i} (\kappa_i h)^2 + \sum_{i} \sum_{\substack{j \\ (i \neq j)}} (f_i h) \left(\frac{\kappa_j^2}{f_j} h \right) \\
&= \int_0^{\Lambda} \mathrm{d}x \int_0^{\Lambda} \left[k_0^2 n^2(x) - \beta^2 \right] \mathrm{d}x \\
&= k_0 \Lambda \int_0^{\Lambda} n^2(x) \mathrm{d}x - \beta^2 \Lambda,
\end{aligned} \tag{3.146}$$

and for the TM mode

$$\begin{aligned}&\sum_{i}(\kappa_i h)^2+\sum_{i}\sum_{\substack{j\\(i\neq j)}}(f_i h)\left(\frac{\kappa_j^2}{f_j}h\right)\\&=\int_0^\Lambda n^2(x)\mathrm{d}x\int_0^\Lambda\frac{k_0^2 n^2(x)-\beta^2}{n^2(x)}\mathrm{d}x\\&=k_0^2\Lambda\int_0^\Lambda n^2(x)\mathrm{d}x-\beta^2\int_0^\Lambda n^2(x)\mathrm{d}x\int_0^\Lambda\frac{1}{n^2(x)}\mathrm{d}x.\end{aligned}\tag{3.147}$$

Based on the Floquet theorem, and applying the thin-film approximation, the trace of the transfer matrix of a period Λ is found to be

$$\begin{aligned}\mathrm{Tr}[M(\Lambda)]&=2\cos K\Lambda\\&=2-(K\Lambda)^2,\end{aligned}\tag{3.148}$$

where K is the transverse wave vector of light in the MQW optical waveguide obtained from the definition in Eq. (3.120). And the core layer effective index of the MQW optical waveguide with arbitrary refractive index distribution can be calculated by Eq. (3.144).

For TE mode, there is

$$\left(n_e^{\mathrm{TE}}\right)^2=\frac{1}{\Lambda}\int_0^\Lambda n^2(x)\mathrm{d}x,\tag{3.149}$$

and for TM mode, it becomes

$$\left(n_e^{\mathrm{TM}}\right)^2=\frac{1}{\Lambda}\int_0^\Lambda n^2(x)dx-\left[\frac{1}{\Lambda^2}\int_0^\Lambda n^2(x)\mathrm{d}x\int_0^\Lambda\frac{1}{n^2(x)}\mathrm{d}x-1\right](\beta/k_0)^2.\tag{3.150}$$

In order to verify the exactness of Eqs. (3.149) and (3.150), let us consider a simple step-index MQW optical waveguide, which is given by

$$n^2(x)=\begin{cases}n_1^2, & 0<x<h_1,\\ n_2^2, & h_1<x<h_1+h_2,\end{cases}\tag{3.151}$$

where $h_1+h_2=\Lambda$. Inserting Eq. (3.151) into Eqs. (3.149) and (3.150), it is easy to derive expressions exactly the same with Eqs. (3.127) and (3.128).

In the actual application, there will be another MQW optical waveguide with refractive index distribution based on trigonometric and hyperbolic functions, so

Eqs. (3.149) and (3.150) are very useful dealing with these problems based on the effective index approximation.

3.4.2 Non-Effective Index Method [11]

The approximation used in the above section is called as the effective index approximation, which uses a constant refractive index to represent the periodic refractive index distribution of the core layer of the MQW optical waveguide. So this method only requires the knowledge of the trace of the transfer matrix corresponds to a period Λ of the MQW optical waveguide. First, the MQW optical waveguide should be approximated as a simple three-layer slab waveguide, which is easy to analyze, but the method is not exact and some physics is also lost. For example, the dispersion in the two media of the core layer cannot be the same. A completely different strategy is adopted in this section, which derives directly the transfer matrix of a period Λ of the MQW optical waveguide, and solves the dispersion equations of the TE and TM polarizations of the MQW optical waveguide with arbitrary refractive index distribution via the boundary condition.

Considering the MQW optical waveguide with arbitrary refractive index as plotted in Fig. 3.13, it is clear that the transfer matrix of a period ε_1 in the MQW optical waveguide is the product of l matrix $M_i (i = 1, 2, \ldots, l)$. And M_i is given by Eqs. (3.140) or (3.141). Different from the last section, it is the transfer matrix $M(\Lambda)$ itself needed in this section, not its trace. The product of the l matrix M_i can be calculated as follows:

$$
\begin{aligned}
M(\Lambda) &= M_1 M_2 \ldots M_{l-1} M_l \\
&= E^l - \sum_i (f_i h)\left(E^{i-1} A E^{l-i}\right) + \sum_i \left(\frac{\kappa_i^2}{f_i} h\right)\left(E^{l-1} B E^{l-i}\right) \\
&\quad - \frac{1}{2}\sum_i (\kappa_i h)^2 E^l + \frac{1}{2}\sum_i \sum_{\substack{j \\ (i \neq j)}} (f_i h)(f_j h)\left(E^{i-1} A E^{j-i-1} A E^{l-j}\right) \\
&\quad + \frac{1}{2}\sum_i \sum_{\substack{j \\ (i \neq j)}} \left(\frac{\kappa_i^2}{f_i} h\right)\left(\frac{\kappa_j^2}{f_j} h\right)\left(E^{i-1} B E^{j-i-1} B E^{l-j}\right) \\
&\quad - \frac{1}{2}\sum_i \sum_{\substack{j \\ (i \neq j)}} (f_i h)\left(\frac{\kappa_j^2}{f_j} h\right)\left(E^{i-1} A E^{j-i-1} B E^{l-j}\right) \\
&\quad - \frac{1}{2}\sum_i \sum_{\substack{j \\ (i \neq j)}} \left(\frac{\kappa_i^2}{f_i} h\right)(f_j h)\left(E^{i-1} B E^{j-i-1} A E^{l-j}\right).
\end{aligned}
\tag{3.152}
$$

The first term on the right side of Eq. (3.152) is the product of the first elements of the l matrix M_i. The second, third, and fourth terms are the sum of all the products, which are obtained by multiplying the $(l-1)$ first elements of the M_i and the second, third, and fourth element of the one matrix left, respectively. Different from Eq. (3.143), the order of the matrix multiplication cannot be varied. The fifth term is calculated by multiplying the $(l-2)$ first element of the M_i and the two second elements of the two matrixes of M_i left and then taking the sum of different i and j. The sixth term is similar with the fifth term, which is calculated by multiplying the $(l-2)$ first element of the M_i and the two third elements of the two matrixes of M_i left and then taking the sum of different i and j. The seventh term uses the product of the $(l-2)$ first elements of M_i and the second and the third elements of the two matrixes left. Switching the two positions of the second and third elements in the multiplication, we can obtain the eighth term. In the above derivation, only the second-order infinitesimal is considered, while all the higher-order terms are ignored.

Based on the characteristics of matrixes E, A, and B, it is clear that

$$
\begin{aligned}
E^l = E &= \begin{bmatrix} 1 & 0 \\ 0 & 1 \end{bmatrix}, \\
E^{i-1}AE^{l-i} = A &= \begin{bmatrix} 0 & 1 \\ 0 & 0 \end{bmatrix}, \\
E^{i-1}BE^{l-i} = B &= \begin{bmatrix} 0 & 0 \\ 1 & 0 \end{bmatrix}, \\
E^{i-1}AE^{j-i-1}AE^{l-j} = A^2 &= \begin{bmatrix} 0 & 0 \\ 0 & 0 \end{bmatrix}, \\
E^{i-1}BE^{j-i-1}BE^{l-j} = B^2 &= \begin{bmatrix} 0 & 0 \\ 0 & 0 \end{bmatrix}, \\
E^{i-1}AE^{j-i-1}BE^{l-j} = AB &= \begin{bmatrix} 1 & 0 \\ 0 & 0 \end{bmatrix}, \\
E^{i-1}BE^{j-i-1}AE^{l-j} = BA &= \begin{bmatrix} 0 & 0 \\ 0 & 1 \end{bmatrix}.
\end{aligned}
$$

So Eq. (3.152) can be rewritten as

$$
\begin{aligned}
M(\Lambda) &= \begin{bmatrix} 1 - \frac{1}{2}\sum\limits_i (\kappa_i h)^2 - \frac{1}{2}\sum\limits_i \sum\limits_{\substack{j \\ (i \neq j)}} (f_i h)\left(\frac{\kappa_j^2}{f_j} h\right) & -\sum\limits_i (f_i h) \\ \sum\limits_i \left(\frac{\kappa_i^2}{f_i} h\right) & 1 - \frac{1}{2}\sum\limits_i (\kappa_i h)^2 - \frac{1}{2}\sum\limits_i \sum\limits_{\substack{j \\ (i \neq j)}} (f_i h)\left(\frac{\kappa_j^2}{f_j} h\right) \end{bmatrix}; \\
&= \begin{bmatrix} 1 - \frac{1}{2}\sum\limits_i \sum\limits_j (f_i h)\left(\frac{\kappa_j^2}{f_j} h\right) & -\sum\limits_i (f_i h) \\ \sum\limits_i \left(\frac{\kappa_i^2}{f_i} h\right) & 1 - \frac{1}{2}\sum\limits_i \sum\limits_j (f_i h)\left(\frac{\kappa_j^2}{f_j} h\right) \end{bmatrix}.
\end{aligned} \tag{3.153}
$$

In Eq. (3.153), the special case of $i = j$ is already included.

When $l \rightarrow \infty$ and $h \rightarrow 0$, the summation in Eq. (3.153) becomes into integration. For TE mode, there is

$$M(\Lambda) = \begin{bmatrix} 1 - \frac{1}{2}\int_0^\Lambda \mathrm{d}x \cdot \int_0^\Lambda [k_0^2 n^2(x) - \beta^2]\mathrm{d}x & -\int_0^\Lambda \mathrm{d}x \\ \int_0^\Lambda [k_0^2 n^2(x) - \beta^2]\mathrm{d}x & 1 - \frac{1}{2}\int_0^\Lambda \mathrm{d}x \cdot \int_0^\Lambda [k_0^2 n^2(x) - \beta^2]\mathrm{d}x \end{bmatrix}, \tag{3.154}$$

and for TM mode, we have

$$M(\Lambda) = \begin{bmatrix} 1 - \frac{1}{2}\int_0^\Lambda n^2(x)\mathrm{d}x \cdot \int_0^\Lambda \frac{k_0^2 n^2(x) - \beta^2}{n^2(x)}\mathrm{d}x & -\int_0^\Lambda \int n^2(x)\mathrm{d}x \\ \int_0^\Lambda \frac{k_0^2 n^2(x) - \beta^2}{n^2(x)}\mathrm{d}x & 1 - \frac{1}{2}\int_0^\Lambda n^2(x)\mathrm{d}x \cdot \int_0^\Lambda \frac{k_0^2 n^2(x) - \beta^2}{n^2(x)}\mathrm{d}x \end{bmatrix}. \tag{3.155}$$

Based on the Floquet theorem, the propagation wave vector K in a periodic structure should obey

$$\cos(K\Lambda) = \frac{1}{2}\mathrm{Tr}[M(\Lambda)],$$

so from Eq. (3.154), we have

$$\cos(K\Lambda) = 1 - \frac{1}{2}\int_0^\Lambda \mathrm{d}x \cdot \int_0^\Lambda \left[k_0^2 n^2(x) - \beta^2\right]\mathrm{d}x. \tag{3.156}$$

In the thin-film approximation, i.e., $K\Lambda \ll 1$, the second term on the right-hand side of the above equation can be treated as a second-order infinitesimal, so

$$\sin(K\Lambda) = \left[1 - \cos^2(K\Lambda)\right]^{1/2} \approx \int_0^\Lambda \mathrm{d}x \cdot \int_0^\Lambda \left[k_0^2 n^2(x) - \beta^2\right]\mathrm{d}x. \tag{3.157}$$

Inserting Eqs. (3.156) and (3.157) into (3.154), the transfer matrix corresponds to a period Λ of the MQW optical waveguide is

$$M(\Lambda) = \begin{bmatrix} \cos(K\Lambda) & -\frac{1}{\eta}\sin(K\Lambda) \\ \eta\sin(K\Lambda) & \cos(K\Lambda) \end{bmatrix}, \tag{3.158}$$

where

$$\eta = \left(\int_0^{\Lambda} [k_0^2 n^2(x) - \beta^2] \mathrm{d}x \Big/ \int_0^{\Lambda} \mathrm{d}x \right)^{1/2}. \tag{3.159}$$

For a core layer of MQW optical waveguide with N periods, its transfer matrix takes the form

$$M(N\Lambda) = \begin{bmatrix} \cos(NK\Lambda) & -\frac{1}{\eta}\sin(NK\Lambda) \\ \eta\sin(NK\Lambda) & \cos(NK\Lambda) \end{bmatrix}. \tag{3.160}$$

For TM mode, the transfer matrixes $M(\Lambda)$ and $M(N\Lambda)$ have the similar forms as Eqs. (3.158) and (3.160), but the variables $K\Lambda$ and η will be defined by

$$\cos(K\Lambda) = 1 - \frac{1}{2} \int_0^{\Lambda} n^2(x)\mathrm{d}x \cdot \int_0^{\Lambda} \frac{k_0^2 n^2(x) - \beta^2}{n^2(x)} \mathrm{d}x, \tag{3.161}$$

$$\eta = \left[\int_0^{\Lambda} \frac{k_0^2 n^2(x) - \beta^2}{n^2(x)} \mathrm{d}x \Big/ \int_0^{\Lambda} n^2(x)\mathrm{d}x \right]^{1/2}. \tag{3.162}$$

In fact, the transfer matrix of a period Λ will be represented by Eqs. (3.154) and (3.155), which are difficult to be extended to the case with several periods. So it is necessary to write Eq. (3.158) by the Floquet theorem first, and then, the transfer matrix of N periods can be obtained via the Chebyshev polynomials, e.g., Equation (3.160).

Using the transfer matrix theory and based on Eqs. (2.44) and (2.45), the dispersion relation of the two polarizations of the MQW optical waveguide can be obtained.

For TE mode, there is

$$\begin{bmatrix} -q & 1 \end{bmatrix} \begin{bmatrix} \cos(NK\Lambda) & -\frac{1}{\eta}\sin(NK\Lambda) \\ \eta\sin(NK\Lambda) & \cos(NK\Lambda) \end{bmatrix} \begin{bmatrix} 1 \\ -p \end{bmatrix} = 0, \tag{3.163}$$

or equivalently

$$\tan(NK\Lambda) = \frac{p+q}{\eta(1 - pq/\eta^2)}, \tag{3.164}$$

where the $K\Lambda$ and η in the two equations above are defined by Eqs. (3.156) and (3.159).

For TM mode, there is

$$\begin{bmatrix} -\frac{1}{n_3^2}q & 1 \end{bmatrix} \begin{bmatrix} \cos(NK\Lambda) & -\frac{1}{\eta}\sin(NK\Lambda) \\ \eta\sin(NK\Lambda) & \cos(NK\Lambda) \end{bmatrix} \begin{bmatrix} 1 \\ -\frac{1}{n_0^2}p \end{bmatrix} = 0. \tag{3.165}$$

or

$$\tan(NK\Lambda) = \frac{p/n_0^2 + q/n_3^2}{\eta\left[1 - pq/(n_0^2 n_3^2 \eta^2)\right]}, \tag{3.166}$$

where the $K\Lambda$ and η are defined by Eqs. (3.161) and (3.162), respectively.

In order to verify these results, consider a step-index MQW optical waveguide made up of GaAs/GaAlAs, and the typical parameters are

$$n_1 = 3.50,\ n_2 = 3.20,\ n_3 = 1.0,\ n_0 = 3.30,\ w = 1\ \mu\text{m},\ h_1 = h_2 = w/2N.$$

The exact dispersion equation is provided by Lenz and Salzman [12]. The results obtained by theory in this section are shown in Fig. 3.14, where the effective refractive index β/k_0 is plotted as a function of the wavelength λ.

As can be seen, when $N = 10$, our result suffers a error of 0.1 % when compared with the exact result. But as the N increases, the accuracy improves significantly.

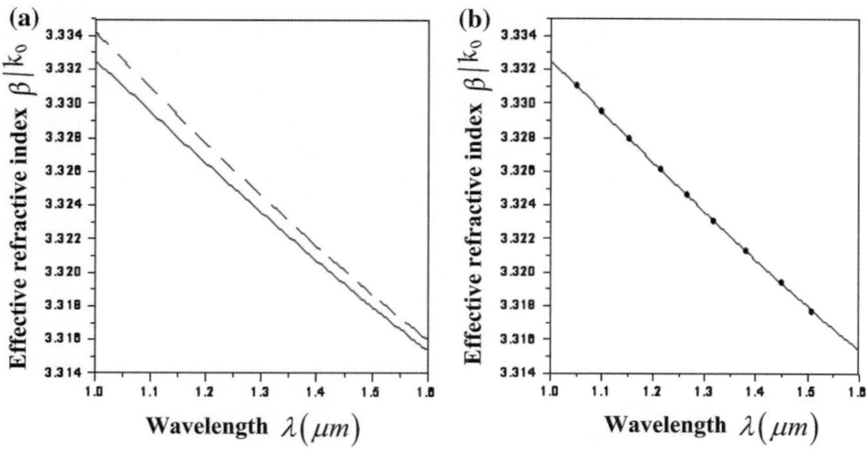

Fig. 3.14 Effective refractive index β/k_0 varies as a function of wavelength λ

When $N = 100$, the two curves coincided with each other. This confirms that the theory of this section can be used under the thin-film approximation, i.e., $K\Lambda \ll 1$.

(a) $N = 10$, and solid and the dashed lines denote the exact and the approximated solution, respectively.
(b) $N = 100$, and solid line is the exact solution, the black dots are the approximated solution.

References

1. A. Yariv, *Quantum Electronics*, Second edn. (Wiley, New York, 1975)
2. Z.Q. Cao, J.X. Fang, Analysis of rectangular-corrugated waveguides by characteristic matrix method. J. Appl. Sci. **2**, 15 (1984). (in Chinese)
3. Z.Q. Cao, J.X. Fang, Coupling coefficients for distributed feedback lasers. J. Appl. Sci. **4**, 36 (1986). (in Chinese)
4. V.M. Agranovich, V.E. Kravtsov, Notes on crystal optics of superlattices. Solid State Commun. **55**, 85 (1985)
5. R.E. Smith, L.A. Molter, M. Dutta, Evaluation of refractive index approximations used for mode determination in multiple quantum well slab waveguides. IEEE J. Quantum Electron. **27**, 1119 (1991)
6. S. Ohke, T. Umeda, Y. Cho, Optical waveguides using GaAs-AlxGa1 − xAs multiple quantum well. Opt. Commun. **56**, 235 (1985)
7. S. Ohke, T. Umeda, Y. Cho, TM-mode propagation and form birefringence in a GaAs-AlGaAs multiple quantum well optical waveguide. Opt. Commun. **70**, 92 (1989)
8. I.M. Skinner, R. Shail, B.L. Weiss, Modal propagation within MQW waveguides. IEEE J. Quantum Electron. **25**, 6 (1989)
9. Z. Cao, New formula of equivalent refractive index in MQW waveguide. Acta Optica Sinica. **11**, 289 (1991). (in Chinese)
10. Z. Cao, C. Hu, and G. Jin, Method of equivalent refractive indices in multi-quantum-well waveguides with arbitrarily shaped base periods. J. Opt. Soc. Am. B **8**, 2519 (1991)
11. Z. Cao, C.N. Ironside, Dispersion relations of multiple quantum well waveguides with arbitrary refractive index profile base period. Opt. Commun. **91**, 203 (1992)
12. G. Lenz, J. Salzman, Eigenmodes of multiwaveguide structures. J. Lightwave Technol. **8**, 1803 (1990)

Chapter 4
Characterizing the Feature Parameters of Planar Optical Waveguide

Abstract In the linear planar optical waveguide, the most interested feature parameters include the guiding layer's refractive index and thickness, the propagation constant, and the propagation loss. On the other side, the feature parameters frequently investigated in the nonlinear planar optical waveguide are the linear and quadratic electro-optical coefficients and the thermo-optical coefficient. Our aim in this chapter was to show the mechanisms and experiment methods of how to exactly characterize these feature parameters, which could lay the groundwork for optimally designing and manufacturing the optical waveguide devices.

Keywords Four-layer leaky waveguide · M-line spectroscopy · Attenuated total reflection spectrum · Inverse WKB method · Inverse analytical matrix method · Propagation loss · Nonlinear optical coefficient

4.1 Four-Layer Leaky Waveguide

Compared to a three-layer planar waveguide (as described in Chap. 2), the four-layer leaky waveguide has an additional layer with a high refractive index (RI) material above its cladding layer. As a result, the guided modes and the radiation modes will couple with each other because the evanescent field produced in the cladding layer can reach to and reflect at the interface between the cladding layer and the added high RI material.

4.1.1 Dispersion Equation

The diagram of light ray propagated in a four-layer leaky waveguide, where the prism is much thicker than the guiding layer and can be regarded as a semi-infinite medium, is shown in Fig. 4.1. Since the RI of prism (n_3) is larger than that of

X. Wang et al., *Progress in Planar Optical Waveguides*,
Springer Tracts in Modern Physics 266, DOI 10.1007/978-3-662-48984-0_4

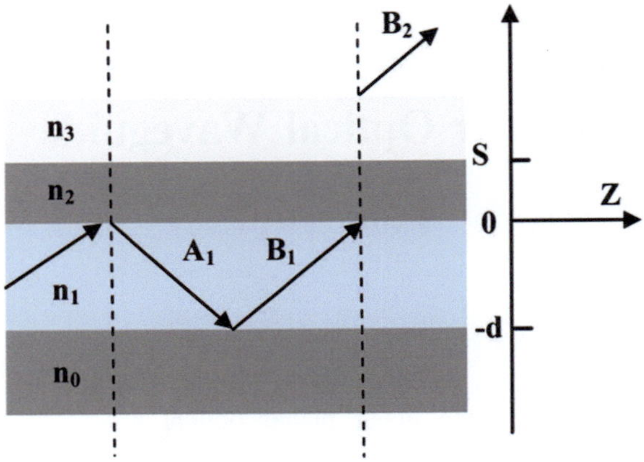

Fig. 4.1 Structure of four-layer leaky waveguide

guiding layer (n_1), the energy of guided modes will continuously radiate into the prism through the air cladding layer (n_2). If the air cladding layer is extended to be large enough, the energy loss radiated from the guided modes will be extremely small and such prism-guide coupling system can be considered as a perturbation model of the three-layer waveguide. Therefore, the dispersion equation of four-layer leaky waveguide can be obtained with a minor modification on that of three-layer waveguide by taking the existence of prism into account. Results indicate that the eigenvalue of guided modes in the four-layer leaky waveguide is not a real but a complex number, which has a real and an imaginary part representing the propagation constant and the attenuation coefficient, respectively.

From bottom to up, we use subscripts 0, 1, 2, and 3 to denote the physical quantities of substrate, guiding layer, air cladding layer, and prism, respectively, and subscripts with two numbers are to represent the physical quantities at the interface between two mediums. Moreover, the thicknesses of guiding layer and air cladding layer are marked as d and s, respectively. The chosen coordinate system is given in Fig. 4.1. Note that the electromagnetic field is uniform in Y-direction and we present only the discussion of TE mode. By setting the propagation factor of electromagnetic field as $\exp[i(\beta z - \omega t)]$, TE mode satisfies the following:

$$\frac{\partial^2 E_y}{\partial x^2} + (k_0^2 n_j^2 - \beta^2)E_y = 0, \quad (j = 0, 1, 2, 3). \tag{4.1}$$

The distribution of electrical field in each layer of the four-layer leaky waveguide is described as follows:

$$E_y(x) = \begin{cases} B_3 \exp[i\kappa_3(x - s)], & (s < x < +\infty), \\ A_2 \exp(p_2 x) + B_2 \exp(-p_2 x), & (0 < x < s), \\ A_1 \exp(-i\kappa_1 x) + B_1 \exp(i\kappa_1 x), & (-d < x < 0), \\ A_0 \exp[p_0(x + d)], & (-\infty < x < -d). \end{cases} \tag{4.2}$$

Substituting Eq. (4.2) into Eq. (4.1), we get the following:

$$\begin{cases} \kappa_3 = (k_0^2 n_3^2 - \beta^2)^{1/2}, \\ \kappa_1 = (k_0^2 n_1^2 - \beta^2)^{1/2}, \\ p_2 = (\beta^2 - k_0^2 n_2^2)^{1/2}, \\ p_0 = (\beta^2 - k_0^2 n_0^2)^{1/2}. \end{cases} \tag{4.3}$$

By applying the boundary continuity conditions at $x = s,\ 0,\ -d$, the transmission type dispersion equation is obtained as follows:

$$\begin{aligned} \exp[i2(\kappa_1 d - \phi_{10} - \phi_{12})] - 1 = \{&\exp[i2(\kappa_1 d - \phi_{10})] - \exp(-2i\phi_{12})\} \\ &\cdot \exp(-2i\phi_{32}) \cdot \exp(-2p_2 s). \end{aligned} \tag{4.4}$$

where

$$\begin{cases} \phi_{10} = \arctan\left(\dfrac{p_0}{\kappa_1}\right), \\ \phi_{12} = \arctan\left(\dfrac{p_2}{\kappa_1}\right), \\ \phi_{32} = \arctan\left(\dfrac{p_2}{\kappa_3}\right). \end{cases} \tag{4.5}$$

If the thickness of air gap is extended to infinite, i.e., $s \to \infty$, Eq. (4.4) will be reduced into the dispersion equation of three-layer waveguide

$$\exp[i2(\kappa_1^{\circ} d - \varphi_{10}^{\circ} - \varphi_{12}^{\circ})] = 1, \tag{4.6}$$

where the superscript "o" represents the physical quantities in the three-layer waveguide.

4.1.2 Variation in the Propagation Constant

In the weak coupling situation, the thickness of air cladding layer s is of the order of light wavelength and stratifies the following inequality

$$\exp(-2p_2 s) \ll 1. \tag{4.7}$$

By utilizing the perturbation theory, the variation in the propagation constant, which is originated from the energy radiation induced by the existence of prism, can be expressed by

$$\Delta\beta^L = -\frac{\kappa_1}{\beta d_{\text{eff}}}\sin 2\phi_{12}\cdot\exp(-i2\phi_{32})\exp(-2p_2 s), \tag{4.8}$$

where

$$d_{\text{eff}} = d + \frac{1}{p_0} + \frac{1}{p_2}. \tag{4.9}$$

In Eqs. (4.8) and (4.9), the superscript "o" representing the physical quantities in three-layer waveguide is omitted for brevity.

The $\Delta\beta^L$ in Eq. (4.8) is a complex number. It means that the existence of prism not only brings an energy loss through radiating (i.e., the imaginary part in the variation of propagation constant), but also leads to a movement of guided modes (i.e., the real part in the variation of propagation constant). Its imaginary part is given as

$$\text{Im}(\Delta\beta^L) = \frac{\kappa_1}{\beta d_{\text{eff}}}\sin 2\phi_{12}\sin 2\phi_{32}\exp(-2p_2 s), \tag{4.10}$$

and real part satisfies

$$\text{Re}(\Delta\beta^L) = -\frac{\kappa_1}{\beta d_{\text{eff}}}\sin 2\phi_{12}\cos 2\phi_{32}\exp(-2p_2 s), \tag{4.11}$$

respectively. Equation (4.10) clearly shows that there is an intimate relationship between the radiative energy loss and the thickness of air cladding layer s, namely the larger s, the smaller radiative energy loss. Consequently, the propagation constant of four-layer leaky waveguide can be expressed as a sum of the propagation constant of three-layer waveguide and a correction

$$\beta^L = \beta^\circ + \Delta\beta^L. \tag{4.12}$$

4.1.3 *Analytical Transfer Matrix Method [1]*

The analytical transfer matrix (ATM) method (described in Chap. 2) is another efficient way to derive the obtained results in Sect. 4.1.2. The matrix equation of four-layer leaky waveguide shown in Fig. 4.1 is written as follows:

$$\begin{bmatrix} E_y(-d) \\ E'_y(-d) \end{bmatrix} = M_1 \cdot M_2 \begin{bmatrix} E_y(s) \\ E'_y(s) \end{bmatrix}, \tag{4.13}$$

where

$$M_1 = \begin{bmatrix} \cos(\kappa_1 d) & -\frac{1}{\kappa_1}\sin(\kappa_1 d) \\ \kappa_1 \sin(\kappa_1 d) & \cos(\kappa_1 d \end{bmatrix}, \tag{4.14}$$

$$M_2 = \begin{bmatrix} \cosh(p_2 s) & -\frac{1}{p_2}\sinh(p_2 s) \\ -p_2 \sinh(p_2 s) & \cosh(p_2 s) \end{bmatrix}. \tag{4.15}$$

Substitution of Eq. (4.2) into Eq. (4.13) yields

$$(-p_0\ 1)\ M_1 M_2 \begin{pmatrix} 1 \\ i\kappa_3 \end{pmatrix} = 0, \tag{4.16}$$

which can be formally rewritten as the phase-type dispersion equation of four-layer leaky waveguide

$$\kappa_1 d = m\pi + \phi_{10} + \phi'_{12}, \quad (m = 0, 1, 2, \ldots.), \tag{4.17}$$

where

$$\phi'_{12} = \arctan(\frac{p'_2}{\kappa_1}), \tag{4.18}$$

and

$$p'_2 = p_2 \frac{\sinh(p_2 s) - \dfrac{i\kappa_3}{p_2}\cosh(p_2 s)}{\cosh(p_2 s) - \dfrac{i\kappa_3}{p_2}\sinh(p_2 s)} = p_2 \frac{1 + \exp(-i2\phi_{32}) \cdot \exp(-2p_2 s)}{1 - \exp(-i2\phi_{32}) \cdot \exp(-2p_2 s)}. \tag{4.19}$$

When the weak coupling condition (i.e., inequality (4.7)) is satisfied, Eq. (4.19) can be approximately expressed as

$$p'_2 \approx p_2[1 + 2\exp(-2i\phi_{32})\exp(-2p_2 s)]. \tag{4.20}$$

Using the differential formula, we obtain

$$\begin{aligned} \varphi'_{12} &= \arctan\left(\frac{p'_2}{\kappa_1}\right) \\ &= \arctan\left(\frac{p_2}{\kappa_1}\right) + \frac{2p_2/\kappa_1}{1 + (p_2/\kappa_1)^2}\exp(-i2\phi_{32}) \cdot \exp(-2p_2 s) \\ &= \varphi_{12} + \sin 2\varphi_{12}\exp(-i2\phi_{32})\exp(-2p_2 s). \end{aligned} \tag{4.21}$$

Therefore, Eq. (4.17) can be modified as

$$\kappa_1 d = m\pi + \phi_{10} + \phi_{12} + \sin 2\phi_{12} \exp(-i2\phi_{32}) \exp(-2p_2 s), \quad (m = 0, 1, 2, \ldots). \tag{4.22}$$

With the application of perturbation theory, the propagation constant of four-layer leaky waveguide β can be expanded in terms of the propagation constant of three-layer waveguide β°, and its first-order Taylor expansion is identical with Eq. (4.8).

4.2 Prism-Waveguide Coupling System

The problem of how to effectively couple the light from the prism with a high RI into a planar waveguide structure is an eternal theme of optics. In 1946, the first experiment for investigating the prism coupling system was reported by Osterberg and Smith [2]. Subsequently, the exciting condition of surface plasmon along the interface between metal film and prism attracted much attention [3]. However, it was not until 1969 that the excitation of guided modes in the dielectric waveguide with a prism coupling structure and a high coupling efficiency was accomplished by a research group in Bell laboratory [4].

4.2.1 Operational Principle and M-Line Spectroscopy [4]

The structure of prism-waveguide coupling system is similar to the four-layer leaky waveguide (see Fig. 4.1). It composes of a high RI prism, a narrow coupling gap (air or metal), a thin film, and a substrate. After passing through the prism, the incident light is totally reflected at the bottom side of the prism and an evanescent field is generated at the interface between prism and coupling gap. Since the thickness of coupling gap is smaller than the decay length of evanescent field, the energy arrived at the interface of coupling gap-thin film can be reflected and another evanescent field with opposite propagation direction is generated. These two evanescent fields interact with each other, and as a result, the incident light from the prism is coupled into the guiding layer (thin film). Note that the coupling process is reversible, namely the energy of guided modes in thin film can also be recoupled into the prism. This energy exchange process between guided modes and radiative modes is called as the optical tunneling effect, which is an optical version of the quantum tunneling effect for particles penetrating through a potential barrier. Mathematically, the scale wave equation for the optical waveguide is quite similar to the Schrödinger equation and the material with a high and a low RI can be regarded as a potential well and a potential barrier, respectively. Therefore, the

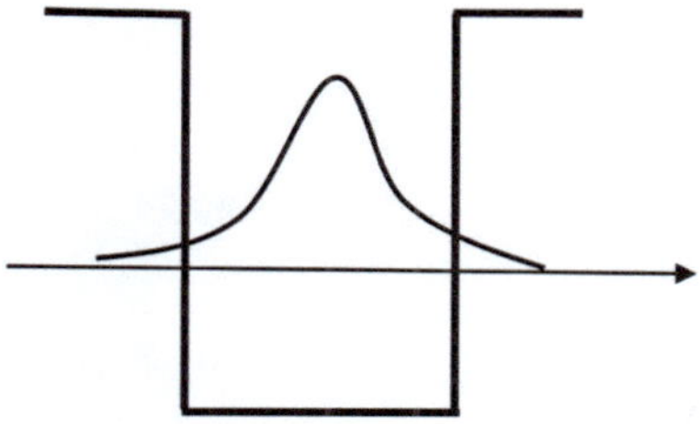

Fig. 4.2 Quantum analog of the three-layer optical waveguide

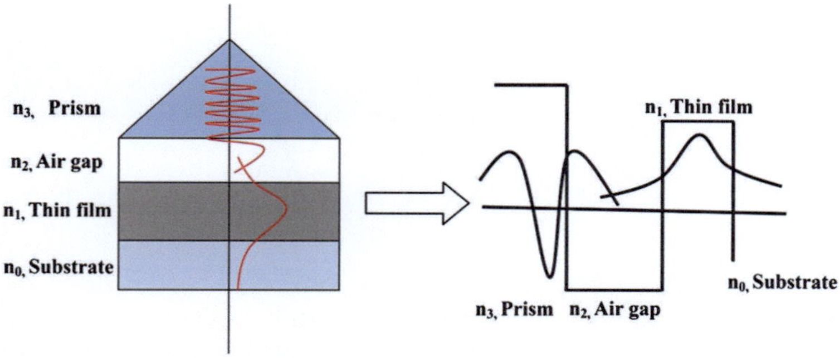

Fig. 4.3 Prism-waveguide coupling system and its RI distribution

three-layer optical waveguide is equivalent to a potential well as shown in Fig. 4.2. Because the potential barriers on both sides of the potential well are semi-infinite, it means that the light energy of optical waveguide can be well localized in the guiding layer.

The RI distribution of the prism-waveguide coupling system is shown in Fig. 4.3. Because the RI of prism is higher than that of guiding layer, the light energies in prism and guiding layer will couple with each other through the optical tunneling effect. Note that such optical tunneling effect only happens when the phase-matching condition is satisfied. It requires that the optical wave vector along z direction in prism should be equal to the propagation constant of guided mode in the guiding layer, i.e.,

$$k_0 n_3 \sin\theta_3 = \beta. \tag{4.23}$$

The above equation is the so-called synchronous condition and θ_3 is the synchronous angle.

The schematic layout of m-line spectrum is shown in Fig. 4.4. The incident light from a side of an isosceles prism (n_3) is coupled into the guiding layer (n_1) and then is recoupled through another side of prism to display the reflected light on a screen. Under the synchronous condition, the energy of incident light will be well coupled into the guiding layer and a guided mode is excited. But in a real guiding layer,

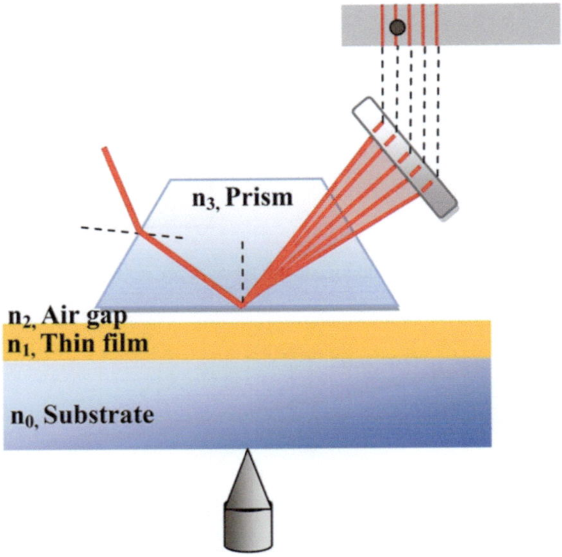

Fig. 4.4 Schematic layout of the m-line spectrum

optical scattering is inevitable and some part of energy in the excited guided mode is likely to be coupled into other adjacent guided modes, which result in several intensity dips at different reflect angles and a set of dark lines emerge on the screen. By measuring these synchronous angles and solving dispersion equation of the prism-waveguide coupling system, both the guiding layer's RI and the thickness can be worked out. The above measuring technique is called as the m-line spectroscopic methodology.

4.2.2 *Reflectivity Formula and Attenuated Total Reflection (ATR) Spectrum*

1. **Reflectivity formula**

The structure parameters and the selected reference frame of the prism-waveguide coupling system are shown in Fig. 4.5. The distribution of electromagnetic field in this system differs somewhat from that of the four-layer leaky waveguide (see Fig. 4.1). That is to say, there is an added incident light ray in the prism-waveguide coupling system. The profile of TE mode excited in the prism-waveguide coupling system is

$$E_y(x) = \begin{cases} A_3 \exp[-i\kappa_3(x-s)] + B_3 \exp[i\kappa_3(x-s)], & (s < x < +\infty), \\ A_2 \exp(-p_2 x) + B_2 \exp(p_2 x), & (0 < x < s), \\ A_1 \exp(-i\kappa_1 x) + B_1 \exp(i\kappa_1 x), & (-d < x < 0), \\ A_0 \exp[p_0(x+d)], & (-\infty < x < -d), \end{cases} \tag{4.24}$$

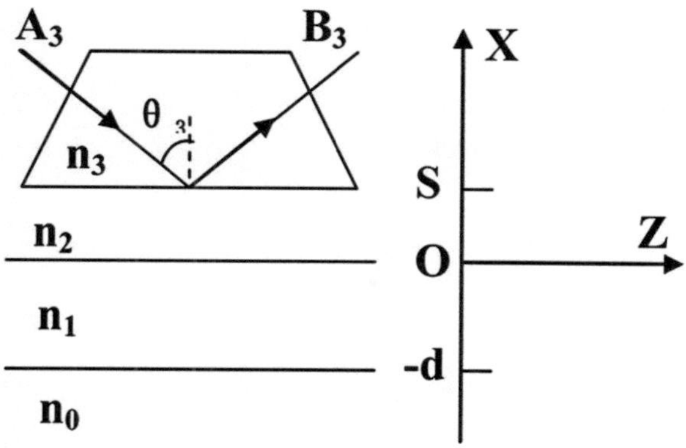

Fig. 4.5 Structure parameters of the prism-waveguide coupling system and the reference frame

where the expressions for κ_1, κ_3, p_0, p_2 are described by Eq. (4.3).

By applying the boundary continuity conditions at interfaces, we get the reflection coefficient

$$r = \frac{B_3}{A_3} = \frac{r_{32} + r_{210}\exp(-2p_2 s)}{1 + r_{32} r_{210}\exp(-2p_2 s)}, \tag{4.25}$$

where

$$\begin{cases} r_{32} = \exp(-i2\phi_{32}), \\ r_{210} = \dfrac{\exp(-i2\phi_{12}) - \exp[i2(\kappa_1 d - \phi_{10})]}{\exp[i2(\kappa_1 d - \phi_{10} - \phi_{12})] - 1}, \end{cases} \tag{4.26}$$

and ϕ_{10}, ϕ_{32}, and ϕ_{12} are given by Eq. (4.5). The propagation constant of the guided mode is determined by the measured synchronous angle

$$\beta = k_0 n_3 \sin\theta_3, \tag{4.27}$$

and the reflectivity is written as

$$R = rr^* = |r|^2. \tag{4.28}$$

2. **Ideal system**

For an ideal system, all the RIs of four medium are real numbers and there is no optical scattering and no optical absorption. It is easy to find that the parameters β, κ_3, κ_1, p_2, and p_0 are real numbers, and consequently ϕ_{10}, ϕ_{12}, and ϕ_{32} are also real numbers. As a result, we obtain

$$r_{32}^{*} = \exp(i2\phi_{32}) = 1/r_{32}, \tag{4.29}$$

$$\begin{aligned} r_{210}^{*} &= \frac{\exp(i2\phi_{12}) - \exp[-i2(\kappa_1 d - \phi_{10})]}{\exp[-i2(\kappa_1 d - \phi_{10} - \phi_{12})] - 1} = \frac{\exp[i2(\kappa_1 d - \phi_{10})] - \exp(-i2\phi_{12})}{1 - \exp[i2(\kappa_1 d - \phi_{10} - \phi_{12})]} \\ &= r_{210}, \end{aligned} \tag{4.30}$$

and

$$r^{*} = \frac{r_{32}^{*} + r_{210}^{*}\exp(-2p_2 s)}{1 + r_{32}^{*} r_{210}^{*}\exp(-2p_2 s)} = \frac{r_{32}^{-1} + r_{210}\exp(-2p_2 s)}{1 + r_{32}^{-1} r_{210}\exp(-2p_2 s)} = \frac{1}{r}. \tag{4.31}$$

It indicates that

$$R = rr^{*} = 1. \tag{4.32}$$

and the incident light in the ideal system cannot be coupled into the guiding layer even when the phase-matching condition is satisfied. In other words, the loss is a necessary condition to cause the energy coupling process in the ideal system.

3. **Reflectivity formula under the phase-matching condition**

However, Eqs. (4.29)–(4.32) are not suitable for the prism-waveguide coupling system because it contains loss (absorption and scattering). If there is no incident light, i.e., $A_3 = 0$, the prism-waveguide coupling system is equivalent to the four-layer leaky waveguide and Eq. (4.25) is reduced into

$$1 + r_{32} r_{210}\exp(-2p_2 s) = 0. \tag{4.33}$$

By substituting Eq. (4.26) into Eq. (4.33), we once again obtain the dispersion Eq. (4.4) of four-layer leaky waveguide. It means that the prism-waveguide coupling system can be regarded as a four-layer leaky waveguide illuminated by an incident light.

Now, let us consider a prism-waveguide coupling system consisting of a lossless prism (i.e., n_3 is real) and a loss guiding layer (i.e., n_1 is complex). The propagation constant β^L is a complex number and expressed as

$$\beta^L = \beta^0 + \Delta\beta^L = \left[\mathrm{Re}(\beta^0) + \mathrm{Re}(\Delta\beta^L)\right] + i\left[\mathrm{Im}(\beta^0) + \mathrm{Im}(\Delta\beta^L)\right], \tag{4.34}$$

where $\mathrm{Im}(\beta^0)$ and $\mathrm{Im}(\Delta\beta^L)$ represent the intrinsic and the radiative dampings [5], respectively. However, the wave vector of incident light in prism is still a real number. As a result, these two wave vectors in prism and guiding layer cannot match well with each other. Only when the imaginary part of β^L is small enough, the phase-matching condition can be approximately satisfied

$$\beta = \mathrm{Re}(\beta^0) + \mathrm{Re}(\Delta\beta^L), \tag{4.35}$$

where β is defined by Eq. (4.27).

The reflection coefficient is

$$\begin{aligned} r &= \frac{r_{32} + r_{210}\exp(-2p_2 s)}{1 + r_{32}r_{210}\exp(-2p_2 s)} = r_{32}\frac{1 + r_{32}^{-1}r_{210}\exp(-2p_2 s)}{1 + r_{32}r_{210}\exp(-2p_2 s)} \\ &= r_{32}\frac{\{\exp[i2(\kappa_1 d - \phi_{10} - \phi_{12})] - 1\} + \{\exp(-2\phi_{12}) - \exp[i2(\kappa_1 d - \phi_{10})]\}r_{32}^{-1}\exp(-2p_2 s)}{\{\exp[i2(\kappa_1 d - \phi_{10} - \phi_{12})] - 1\} + \{\exp(-2\phi_{12}) - \exp[i2(\kappa_1 d - \phi_{10})]\}r_{32}\exp(-2p_2 s)}. \end{aligned} \tag{4.36}$$

Under the conditions of weak coupling ($\exp(-2p_2 s) \ll 1$), we replace β with its Taylor's expansion around β^0 (the propagation constant of three-layer waveguide) and obtain

$$r = r_{32}\frac{(\beta - \beta^0) - \dfrac{i\kappa_1}{2\beta d_{eff}}\{\exp[i2(\kappa_1 d - \phi_{10}) - \exp(-i2\phi_{12})]\}r_{32}^{-1}\exp(-2p_2 s)}{(\beta - \beta^0) - \dfrac{i\kappa_1}{2\beta d_{eff}}\{\exp[i2(\kappa_1 d - \phi_{10}) - \exp(-i2\phi_{12})]\}r_{32}\exp(-2p_2 s)}. \tag{4.37}$$

It is easy to see that the second term in the denominator of Eq. (4.37) equals to $\Delta\beta^L$ (see Eq. 4.8). Due to $\mathrm{Re}(r_{32}^{-1}) = \mathrm{Re}(r_{32})$ and $\mathrm{Im}(r_{32}^{-1}) = -\mathrm{Im}(r_{32})$, Eq. (4.37) is rewritten as

$$r = r_{32}\frac{\beta - [\mathrm{Re}(\beta^0) + \mathrm{Re}(\Delta\beta^L)] - i[\mathrm{Im}(\beta^0) - \mathrm{Im}(\Delta\beta^L)]}{\beta - [\mathrm{Re}(\beta^0) + \mathrm{Re}(\Delta\beta^L)] - i[\mathrm{Im}(\beta^0) + \mathrm{Im}(\Delta\beta^L)]}, \tag{4.38}$$

and the reflectivity is recast into

$$\begin{aligned} R &= |r_{32}|^2\frac{\{\beta - [\mathrm{Re}(\beta^0) + \mathrm{Re}(\Delta\beta^L)]\}^2 + [\mathrm{Im}(\beta^0) - \mathrm{Im}(\Delta\beta^L)]^2}{\{\beta - [\mathrm{Re}(\beta^0) + \mathrm{Re}(\Delta\beta^L)]\}^2 + [\mathrm{Im}(\beta^0) + \mathrm{Im}(\Delta\beta^L)]^2} \\ &= |r_{32}|^2\left(1 - \frac{4\mathrm{Im}(\beta^0)\mathrm{Im}(\Delta\beta^L)}{\{\beta - [\mathrm{Re}(\beta^0) + \mathrm{Re}(\Delta\beta^L)]\}^2 + [\mathrm{Im}(\beta^0) + \mathrm{Im}(\Delta\beta^L)]^2}\right), \end{aligned} \tag{4.39}$$

respectively. Under the approximate phase-matching condition (Eq. 4.35), the reflectivity approaches its minimum, i.e.,

$$R_{\min} = |r_{32}|^2\left\{1 - \frac{4\mathrm{Im}(\beta^0)\mathrm{Im}(\Delta\beta^L)}{[\mathrm{Im}(\beta^0) + \mathrm{Im}(\Delta\beta^L)]^2}\right\}. \tag{4.40}$$

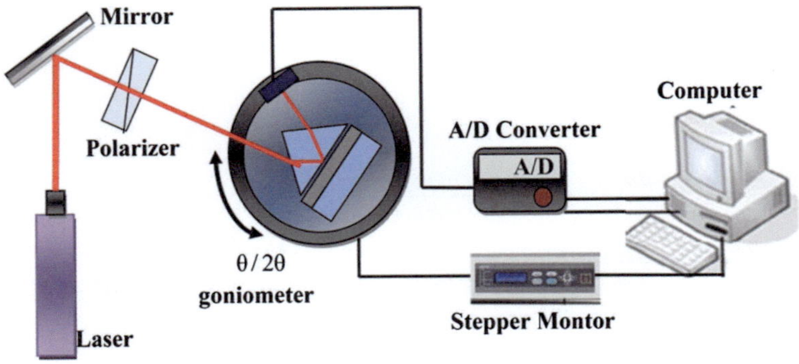

Fig. 4.6 Setup for measuring the ATR spectrum

Moreover, if the intrinsic damping and the radiative damping are equal to each other

$$\mathrm{Im}\left(\beta^0\right) = \mathrm{Im}\left(\Delta\beta^L\right), \tag{4.41}$$

the minimum of reflectivity calculated from Eq. (4.40) is zero

$$R_{\min} = 0. \tag{4.42}$$

4.2.3 *Measuring the Waveguide Layer's Thickness and RI*

The experimental setup for measuring the waveguide layer's thickness and RI is shown in Fig. 4.6. It consists of a semiconductor laser, a mirror, a polarizer, a detector, and a computer-controlled $\theta/2\theta$ goniometer. By scanning the incident angle θ_3, a series of ATR intensity dips will appear in the reflection spectrum and each of them occurs at the corresponding synchronous angle.

A typical ATR spectrum calculated from the reflectivity formula as a function of incident angle is plotted in Fig. 4.7. The dispersion equation is approximated by

$$\kappa_1 h = m\pi + \arctan\left(\frac{f_1}{f_2}\frac{p_2}{\kappa_1}\right) + \arctan\left(\frac{f_1}{f_3}\frac{p_3}{\kappa_1}\right), \quad m = 0, 1, 2, \ldots, \tag{4.43}$$

where

$$\begin{cases} \kappa_1 = \left(k_0^2 n_1^2 - \beta^2\right)^{1/2} \\ p_2 = \left(\beta^2 - k_0^2 n_2^2\right)^{1/2} \\ p_3 = \left(\beta^2 - k_0^2 n_3^2\right)^{1/2} \end{cases}, \tag{4.44}$$

and

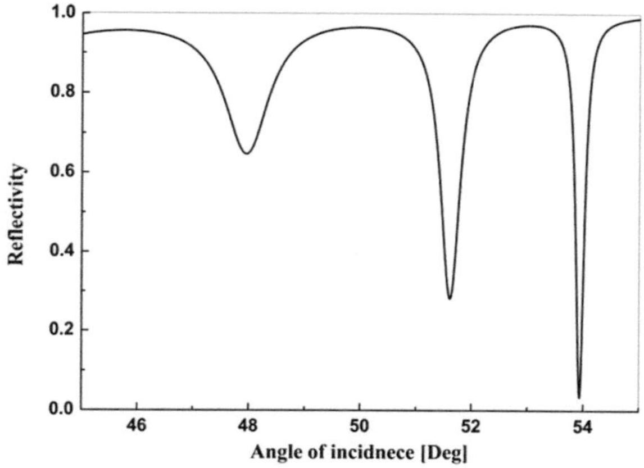

Fig. 4.7 A typical ATR spectrum

$$f_j = \begin{cases} 1 & \text{For TE mode} \\ n_j^2 & \text{For TM mode} \end{cases} \quad (j = 0, 1, 2), \tag{4.45}$$

β is the propagation constant, $k_0 = 2\pi/\lambda$ is the wave vector in vacuum, and λ is the wavelength of incident light. n_1 and n_2 are the RI of guiding layer and substrate (air), n_3 is the complex RI of the cladding layer (silver film), and h is the thickness of guiding layer. The effective RI of guided mode is defined as

$$n_{\text{eff}} = \beta/k_0 = n_{\text{prism}} \sin\theta, \tag{4.46}$$

and the propagation constant β can be easily determined by the measurement of synchronous angle θ. After the determination of propagation constants of three adjacent guided modes (β_{m-1}, β_m, β_{m+1}), two feature parameters, i.e., the guiding layer's thickness (h) and RI (n_1) can be worked out through solving the transcendental equations

$$\begin{cases} \kappa_{m-1}h = (m-1)\pi + \arctan\left(\frac{n_1^2}{n_2^2}\frac{p_{m-1}}{\kappa_{m-1}}\right) + \arctan\left(\frac{n_1^2}{n_3^2}\frac{q_{m-1}}{\kappa_{m-1}}\right), \\ \kappa_m h = m\pi + \arctan\left(\frac{n_1^2}{n_2^2}\frac{p_m}{\kappa_m}\right) + \arctan\left(\frac{n_1^2}{n_3^2}\frac{q_m}{\kappa_m}\right), \\ \kappa_{m+1}h = (m+1)\pi + \arctan\left(\frac{n_1^2}{n_2^2}\frac{p_{m+1}}{\kappa_{m+1}}\right) + \arctan\left(\frac{n_1^2}{n_3^2}\frac{q_{m+1}}{\kappa_{m+1}}\right). \end{cases} \tag{4.47}$$

In the above equation, the thickness (s) of coupling layer is assumed to be semi-infinite, but the actual thickness (s) is a finite value; therefore, the guiding layer's thickness (h) and RI (n_1) should be calculated by Eq. (4.4) with a known s. However, there is a great difficulty in the accurate measurement of s. But in the situation of weak coupling, the influence of prism can be ignored and thus the guiding layer's thickness (h) and RI (n_1) derived from Eq. (4.47) are accurate.

4.3 Determining the RI Profile of Inhomogeneous Waveguide

The waveguide's RI profile can give useful information about the waveguide's fabrication process and transmission properties. So, its determination is of fundamental importance. Given a set of mode indexes n_m, $m = 1, 2, \ldots, M$, what is waveguide's RI profile $n(x)$? Basically, this is an inversion of the Sturm–Liouville eigenvalue problem [6]. Since the given amount of information is finite, one obviously cannot predict the continuous function $n(x)$ exactly. Instead, many approximate methods have been developed. In this subsection, two of them, i.e., inverse WKB method and inverse ATM method, are presented.

4.3.1 Inverse WKB Method [7]

Consider a step-asymmetrical RI profile, which is continuous and monotonically decreasing for $x \geq 0$ with a peak RI n_f at the surface boundary, shown in Fig. 4.8. n_0 and n_s are the RI of the cladding layer and the substrate, respectively. Profiles of this kind are typical in the diffused optical waveguides.

The WKB approximate dispersion equation is

$$k_0 \int_0^{x_m} [n^2(x) - n_m^2]^{1/2} \mathrm{d}x = m\pi + \tan^{-1}\left(\frac{P_0}{\kappa}\right) + \frac{\pi}{4}, (m = 0, 1, 2, \ldots, M-1), \tag{4.48}$$

where $P_0 = \left(\beta^2 - k_0^2 n_0^2\right)^{1/2}$, $\kappa = \left(k_0^2 n_f^2 - \beta^2\right)^{1/2}$, the mode index (effective RI) is $n_m = \beta / k_0$, $k_0 = 2\pi/\lambda$ is the wave number in the vacuum, and the turning point x_m is determined by $n(x_m) = n_m$. In the actual waveguide, n_f is closer to n_s, their difference is small, but the difference between n_f and n_0 is relatively large; therefore, the

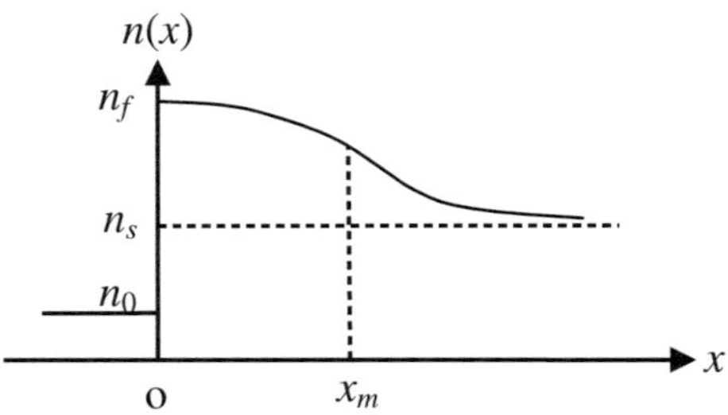

Fig. 4.8 RI profile of an inhomogeneous optical waveguide

second term in the right side of Eq. (4.48) can be approximated as $\pi/2$. By normalizing the variable x to the free-space wavelength λ, Eq. (4.48) is rewritten as

$$\int_0^{x_m} [n^2(x) - n_m^2]^{1/2} \mathrm{d}x = \frac{4m-1}{8}, \quad (m = 1, 2, \ldots, M), \tag{4.49}$$

where the mode order m starts with 1 not the commonly used 0.

Since $n_m (m = 1, 2, \ldots, M)$ can be measured by using the m-line technology, namely the position of n_m at vertical axis $n(x)$ can be determined. To determine the values of x_m from the values of n_m, one can proceed by writing Eq. (4.49) as a sum of integrals

$$\sum_{l=1}^{m} \int_{x_{l-1}}^{x_l} [n^2(x) - n_m^2]^{1/2} \mathrm{d}x = \frac{4m-1}{8}. \tag{4.50}$$

Next, we assume that $n(x)$ is a piecewise linear function connecting the measured values of n_m,

$$n(x) \approx n_l + \frac{n_{l-1} - n_l}{x_l - x_{l-1}} (x_l - x), (x_{l-1} \leq x \leq x_l). \tag{4.51}$$

Let the integral variable in Eq. (4.50) is modified as

$$\left[n^2(x) - n_m^2\right]^{1/2} = [n(x) + n_m]^{1/2} \cdot [n(x) - n_m]^{1/2},$$

and $n(x) + n_m$ is replaced by a midpoint value of $[(n_{l-1} + n_l)/2] + n_m$ for $(x_{l-1} \leq x \leq x_l)$, Eq. (4.50) can be rewritten as follows:

$$\sum_{l=1}^{m} \left(\frac{n_{l-1} + n_l}{2} + n_m\right)^{1/2} \int_{x_{l-1}}^{x_l} \left[(n_l - n_m) + \frac{n_{l-1} - n_l}{x_l - x_{l-1}} (x_l - x)\right]^{1/2} \mathrm{d}x = \frac{4m-1}{8}. \tag{4.52}$$

The solution for x_m is given as

$$x_m = x_{m-1} + \left[\frac{3}{2}\left(\frac{n_{m-1} + 3n_m}{2}\right)^{-1/2} \cdot (n_{m-1} - n_m)^{-1/2}\right]$$
$$\times \left\{\frac{4m-1}{8} - \frac{2}{3} \sum_{l=1}^{m-1} \left(\frac{n_{l-1} + n_l}{2} + n_m\right)^{1/2} \left(\frac{x_l - x_{l-1}}{n_{l-1} - n_l}\right) \cdot \left[(n_{l-1} - n_m)^{3/2} - (n_l - n_m)^{3/2}\right]\right\},$$
$$(m = 2, 3, \ldots, M), \tag{4.53}$$

and

$$x_1 = \frac{9}{16}\left(\frac{n_f + 3n_1}{2}\right)^{-1/2}\left(n_f - n_1\right)^{-1/2}. \tag{4.54}$$

Therefore, we get a simple algorithm for calculating $x_1, x_2, \ldots, x_M$ based on the measured values of $n_1, n_2, \ldots, n_m$. Finally, $n(x)$ is determined approximately by Eq. (4.51).

In the inverse WKB method, the surface index n_f is unknown. This could result in a set of equations with one more freedom degree than the number of measured parameters. The value of n_f mainly affects the low-order modes and does not affect the higher order modes. The value of n_f is always derived from the minimizing the sum of the squares of the second differences of the curve [7] and the function extrapolation [8].

4.3.2 *Inverse ATM Method [9]*

The dispersion equation derived by the ATM method is given as

$$k_0 \int_0^{x_m} \left[n^2(x) - n_m^2\right]^{1/2} \mathrm{d}x + \Phi(s) = (m-1)\pi + \arctan\left(\frac{P_0}{\kappa_f}\right) + \arctan\left(\frac{P_t}{\kappa_l}\right),$$
$$(m = 1, 2, \ldots, M), \tag{4.55}$$

and all used parameters have been described in Chap. 2. The model of RI profile shown in Fig. 4.8 can be depicted as

$$n(x) = \begin{cases} n_l + \dfrac{n_{l-1} - n_l}{x_l - x_{l-1}}(x_l - x) & x_{l-1} \le x \le x_l, \\ n_s + b\exp(-\alpha x) & x > x_M, \end{cases} \tag{4.56}$$

where b and α can be easily determined by the continuous condition at x_M.

With the assumed model of $n(x)$ above, the RI profile can be constructed by the following steps:

(i) Using the inverse WKB method to determine the surface index n_f with the measured mode indexes. Then assuming an initial set of n_m denoted as $\{n_m\}^{(0)}$. Choose of $\{n_m\}^{(0)}$ is quite flexible. $\{n_m\}^{(0)}$ can be an arbitrary increasing series as long as enough modes can be existed. For example, $\{n_m\}^{(0)}$ can be chosen as an equal-spaced increasing series. Then, we obtain an initial $\{n(x)\}^{(0)}$ by substituting the $\{n_m\}^{(0)}$ into Eq. (4.56).

(ii) Substituting the $\{n(x)\}^{(0)}$ into the Eq. (4.55), a new set of $\{x_m\}^{(1)}$ is obtained. Use the $\{x_m\}^{(1)}$ to get a new RI profiles $\{n(x)\}^{(1)}$ from Eq. (4.56). Then calculate the deviation $\delta = \sum_{m=1}^{M} (n_m - \overline{n_m})^2$.

(iii) Repeat step (ii) till the deviation $\delta \to 0$, then the RI profile of waveguide is finally obtained.

After such iterations, δ will be close to zero, which represents that the mode indexes allowed by the waveguide with the obtained RI profile are almost equal to those got from experiment. So, the inverse ATM method is self-consistent.

To illustrate the reliability of inverse ATM method, we take sets of mode indexes calculated exactly by the ATM method from the exponential, Fermi, and step RI profile distributions to predict the real RI profile, and compare the results obtained by the inverse WKB method. In the following calculations, we choose air as the cladding layer $n_0 = 1.0$ and the light wavelength is 632.8 nm. Figure 4.9 indicates that the inverse ATM method is reliable not only in the slowly changing RI profiles but also in the rapidly changing RI profiles, in which the inverse WKB method would produce an unacceptable calculation error.

4.4 Measuring the Waveguide's Propagation Loss

The propagation loss, which is a vital parameter for evaluating the guiding layer, can be classified into three categories:

(i) The scattering loss arising from the inhomogeneous RI distribution and the roughness along the waveguide interfaces;

(ii) The absorption loss originating from the material ions located at the crystal lattice, and the impurity ions or electrons;

(iii) The radiative loss owing to the bend of waveguide and the optical tunneling effect.

The above three categories of losses are described by a uniform parameter: the attention coefficient $\alpha(\text{cm}^{-1})$ or the propagation loss $\xi(\text{dB/cm})$. Supposing that the light is guided from z_1 to z_2 and its intensity decreases from p_1 to p_2, the uniform parameters α and ξ are defined as

$$\alpha = -\frac{1}{z_2 - z_1} \ln\left(\frac{p_2}{p_1}\right), \tag{4.57}$$

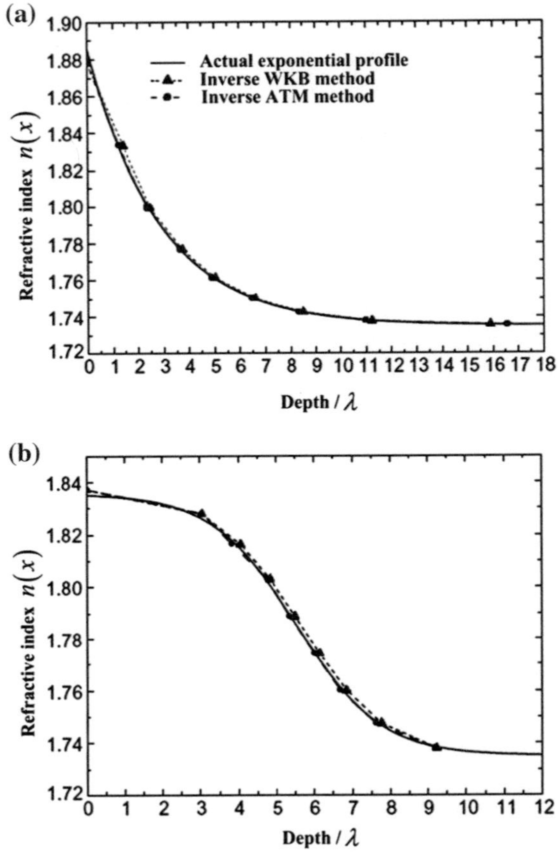

Fig. 4.9 Recovery of the RI profile by the inverse ATM method and the inverse WKB method. **a** The exponential profile, $n(x) = 1.735 + 0.150\exp(-x/1.8), (x, \mu\text{m})$. **b** The slowly changing Fermi function profile, $n(x) = 1.735 + 0.100/\{1 - \exp(-5.0) + \exp[(x - 3.5)/0.7]\}, (x, \mu\text{m})$. **c** The steeper changing Fermi function profile, $n(x) = 1.735 + 0.100/\{1 - \exp(-25.0) + \exp[(x - 2.5)/0.1]\}, (x, \mu\text{m})$,. **d** Step function profile

and

$$\xi = -\frac{10}{z_2 - z_1}\lg\left(\frac{p_2}{p_1}\right), \tag{4.58}$$

respectively.

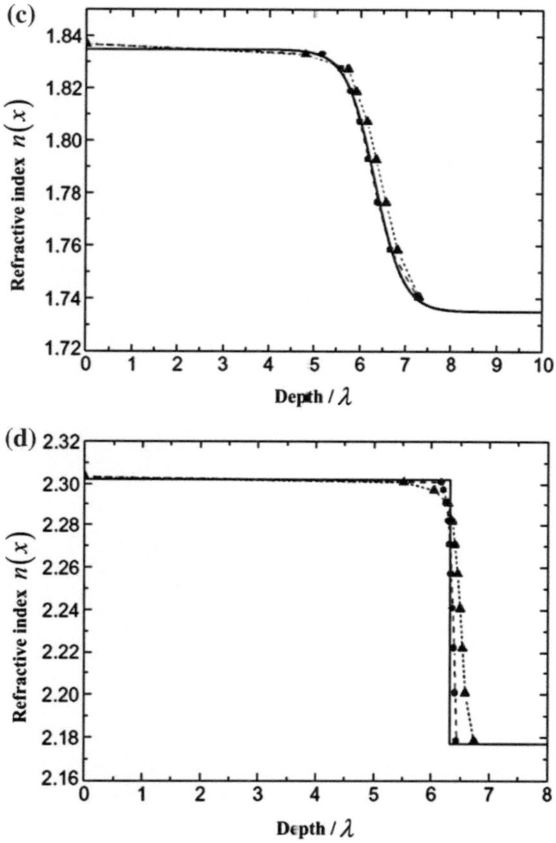

Fig. 4.9 (continued)

4.4.1 Perturbation Analysis of the Propagation Loss

For calculating the propagation loss of optical waveguide, several numerical methods can provide a high-precision result but are unable to give an explicit analytical expression, and thereby several approximate methods have been developed. The physical picture of the frequently used approximate method, i.e., the ray method, is quite clear but its calculation error is always beyond the tolerable level. Here, we describe the perturbation method in detail. As an example, we set the guiding layer as an absorptive material and regard a waveguide with loss (characterizing by a complex RI) as a perturbation system of the ideal waveguide (characterizing by a real RI).

A waveguide containing an absorptive guiding layer is shown in Fig. 4.10. The RI of substrate (n_0) and cladding layer (n_2) are real, but the guiding layer's RI is a complex number

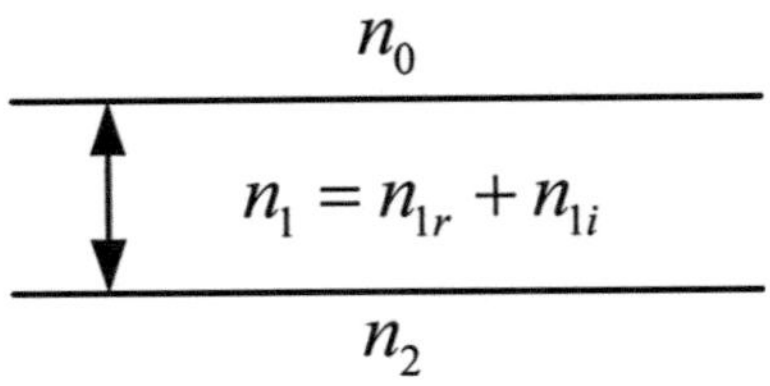

Fig. 4.10 The waveguide containing an absorptive guiding layer

$$n_1 = n_{1r} + in_{1i}, \tag{4.59}$$

and

$$n_{1i} \ll n_{1r}. \tag{4.60}$$

The dispersion equation is

$$\kappa_1 h = m\pi + \arctan\left(\frac{p_0}{\kappa_1}\right) + \arctan\left(\frac{p_2}{\kappa_1}\right), \tag{4.61}$$

where p_0 and p_2 are described by Eq. (4.3), and κ_1 is

$$\kappa_1 = \left(k_0^2 n_1^2 - \beta^2\right)^{1/2} \approx \kappa_{1r} + i\kappa_{1i}, \tag{4.62}$$

where

$$\begin{cases} \kappa_{1r} = \sqrt{k_0^2 n_{1r}^2 - \beta^2}, \\ \kappa_{1i} = \dfrac{k_0^2 n_{1r} n_{1i}}{\sqrt{k_0^2 n_{1r}^2 - \beta^2}}. \end{cases} \tag{4.63}$$

Only keeping the first-order perturbation term, we obtain

$$\begin{cases} \arctan\left(\dfrac{p_0}{\kappa_1}\right) = \arctan\left(\dfrac{p_0}{\kappa_{1r} + i\kappa_{1i}}\right) = \arctan\left(\dfrac{p_0}{\kappa_{1r}}\right) - i\dfrac{p_0\kappa_{1i}}{\kappa_{1r}^2 + p_0^2}, \\ \arctan\left(\dfrac{p_2}{\kappa_1}\right) = \arctan\left(\dfrac{p_2}{\kappa_{1r} + i\kappa_{1i}}\right) = \arctan\left(\dfrac{p_2}{\kappa_{1r}}\right) - i\dfrac{p_2\kappa_{1i}}{\kappa_{1r}^2 + p_2^2}, \end{cases} \tag{4.64}$$

and get the rewritten dispersion equation

$$\begin{aligned} \kappa_{1r} h = m\pi &+ \arctan\left(\frac{p_0}{\kappa_{1r}}\right) + \arctan\left(\frac{p_2}{\kappa_{1r}}\right) \\ &- i\kappa_{1i}\left(h + \frac{p_0}{\kappa_{1r}^2 + p_0^2} + \frac{p_2}{\kappa_{1r}^2 + p_2^2}\right), \quad (m = 0, 1, 2 \ldots). \end{aligned} \tag{4.65}$$

We define a new variable as

$$h_{\text{loss}} = h + \frac{p_0}{\kappa_{1r}^2 + p_0^2} + \frac{p_2}{\kappa_{1r}^2 + p_2^2}, \tag{4.66}$$

and it denotes the transverse attenuation length. With a close inspection on Eq. (4.65), we find that the parameter κ_{1i} represents the transverse attenuation coefficient.

The perturbation of propagation constant ($\Delta\beta$) between the waveguide with loss (β) and the ideal waveguide (β^0) is expressed as

$$\Delta\beta = i\frac{k_0^2 n_{1r} n_{1i}}{\beta} \frac{h + \frac{p_0}{\kappa_{1r}^2 + p_0^2} + \frac{p_2}{\kappa_{1r}^2 + p_2^2}}{h + \frac{1}{p_0} + \frac{1}{p_2}} = i\frac{k_0^2 n_{1r} n_{1i}}{\beta} \frac{h_{\text{loss}}}{h_{\text{eff}}}. \tag{4.67}$$

For brevity, the superscript "o" in the above equation has been omitted.

On the basis of Eq. (4.34), we obtain

$$\Delta\beta = i\text{Im}(\beta). \tag{4.68}$$

The first-order perturbation theory indicates that the absorption in the guiding layer leads to an energy loss but is not likely to change other waveguide's characteristics.

The propagation loss calculated by the perturbation theory and the exact numerical method for the TE_0 mode excited in an actual waveguide is shown in Fig. 4.11. The simulated waveguide parameters are as follows: $n_{1r} = 1.56$, $n_0 = 1.49$, $n_2 = 1.33$, $h = 0.4\mu\text{m}$, and $\lambda = 0.55\mu\text{m}$. The perturbation method agrees well with the exact numerical method if the extinction coefficient (n_{1i}) is smaller than 0.01. Therefore, we can conclude that the utilization of perturbation theory to calculate the propagation loss of waveguide is fairly reliable since the case of the extinction coefficient being larger than 10^{-3} is rare.

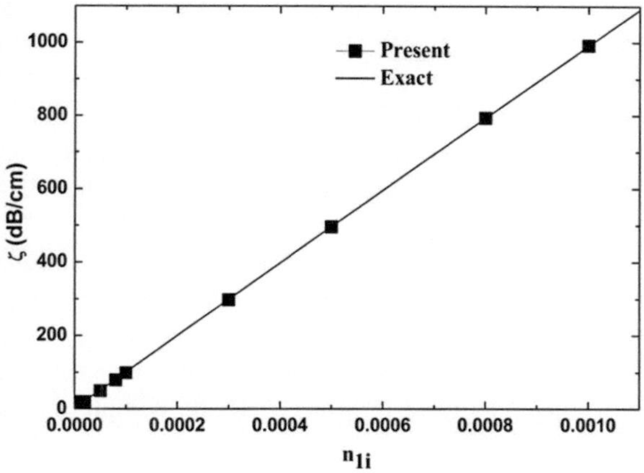

Fig. 4.11 Propagation loss as a function of the extinction coefficient

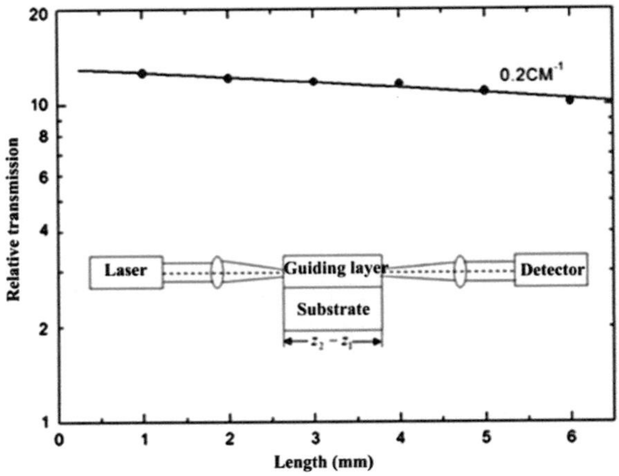

Fig. 4.12 End-face coupling method for measuring the propagation loss of waveguide

4.4.2 End-Face Coupling Method [10]

The schematic diagram for the end-face coupling method to measure the propagation loss of waveguide is shown in Fig. 4.12. A laser beam passes through a convex lens to be focused at one end face of the waveguide and then be coupled into the guiding layer, and finally comes out from another end-face. The input power (P_1) and the output power (P_2) at two end-faces are measured and the propagation loss of waveguide is determined by Eq. (4.57) or Eq. (4.58). However, because the coupling process also leads to energy loss, one has to repeatedly measure a series of identical waveguides with different lengths to eliminate the coupling induced influence. The method starts with a long waveguide, whose length is gradually decreased by cutting down the end face and polishing the newly formed end face after each measurement. To optimize the coupling process and maximize the output power, the laser beam should be well aligned and focused on the same point at the end face of waveguide. All attenuation coefficients from the multiple measurements are located around a straight line in the logarithmic coordinate, and its degree of scatter depends on the uniformity of coupling loss, namely relays on the identity of the end-face polishing quality and the aligning state between laser beam and end face. If these data points are well fitted by a straight line, it indicates that the coupling condition for each measurement is almost the same and the influence of coupling loss can be eliminated by employing the multiple measurements.

The end-face coupling method has three-technique drawbacks to be conquered. First, the guided light beam through a long waveguide would like to spread along the guiding layer and cannot be well focused by a convex lens at the out end face. This divergence of light beam inevitably causes a measurement error. Second, the polishing quality of end face must be completely identical at each measurement,

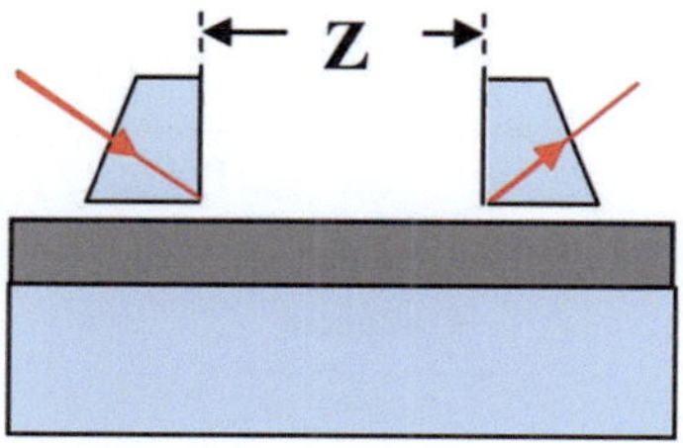

Fig. 4.13 The sliding-prism method to measure the propagation loss of waveguide

which is a near impossibility. Third, the requirement of the same alignment state for each measurement is quite difficult to realize.

4.4.3 Sliding-Prism Method [11]

The setup of the sliding-prism method to measure the propagation loss of waveguide is depicted in Fig. 4.13. The light is coupled into the guiding layer by a fixed prism and then coupled out from the guiding layer by a slideable prism, which can be mechanically slid along the waveguide since a small amount of refractive index-matching oil (its RI is equal to or slightly smaller than that of prism) is inserted between prism and waveguide. The propagation loss of waveguide can be worked out by the dependence of output power on the distance between these two prisms. Compared to the end-face coupling method, the sliding-prism method has several merits, such as the nondestructiveness test and the elimination of end-face polishing. However, a high-precision measurement of propagation loss is difficult because the gap between prism and waveguide is varied and the out-coupling coefficient is inconstant during the output prism sliding. To eliminate the inaccuracy induced by the inconstant out-coupling coefficient, a propagation loss measurement method with three prisms has been proposed. Unfortunately, it needs a longer waveguide and a more complex optical arrangement.

4.4.4 Digital Scattering Method [12]

As described in Sect. 4.2.2, there is no scattering light in an ideal planar waveguide. But part of guided mode energy would be scattered out from the actual waveguide owing to the existence of interface roughness and impurity. The intensity of scattered light is mainly determined by three parameters: the intensity of propagation light, the degree of interface roughness, and the quantity of impurities. Generally, the impurities are randomly distributed in the waveguide. Even though the strong scattered light appears at certain points because of the impurities fluctuation, it can

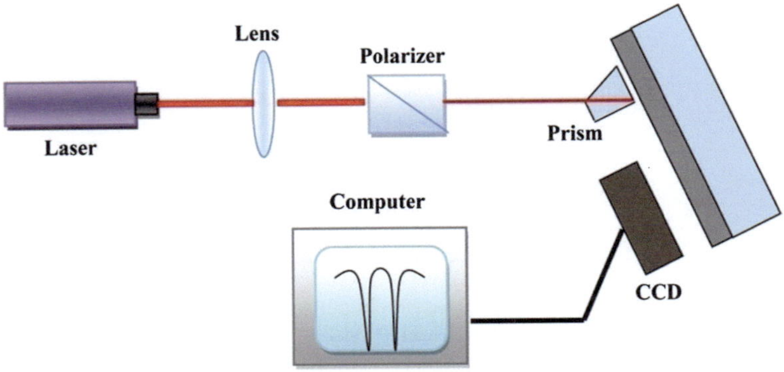

Fig. 4.14 The configuration of digital scattering method

be effectively eliminated by the aid of digital filtering technology. As a result, the intensity of light scattered from each point is proportional to its local intensity of propagation light. In the experiment, the intensity of scattered light along the propagation path is recorded by a CCD camera and is linearly converted into the local intensity of propagation light. Finally, the propagation loss is determined by fitting the local intensity curve of propagation light.

The configuration of digital scattering method is illustrated in Fig. 4.14. The CCD camera is placed further than 5 cm away from the waveguide to enable that the record over the entire waveguide can be done in one shooting. Therefore, it can minimize the disturbance of probe on guided modes and can avoid the measurement error in the multiple shootings method, where the distance between probe and waveguide is difficult to keep unchanged. A photograph of one waveguide with a polymer guiding layer and the recorded scattered light intensity along the propagation path are given in Fig. 4.15. The noise signals in Fig. 4.15b can be eliminated by using the median filtering algorithm, which assumes that the noise signal is random and independent on the intensities of adjacent points. Consequently, the isolated noise signal at certain point should be removed while the median value of all pixels located in a selected window is added to this point.

The median filtering algorithm is one kind of spatial convolution methods, and each data performs the following convolution calculation

$$F(x,y) = \sum_m \sum_n f(x-m, y-n)W(m,n), \tag{4.69}$$

where $f(x-m, y-n)$ and $F(x,y)$ are the image data before and after the convolution calculation, respectively, $W(m,n)$ is a window function to evaluate the median value. The photograph of waveguide and the intensity of scattered light after the spatial convolution operation are shown in Fig. 4.16. With comparison to Fig. 4.15, all the sharp noise signals have been effectively removed at the cost of reduced image details.

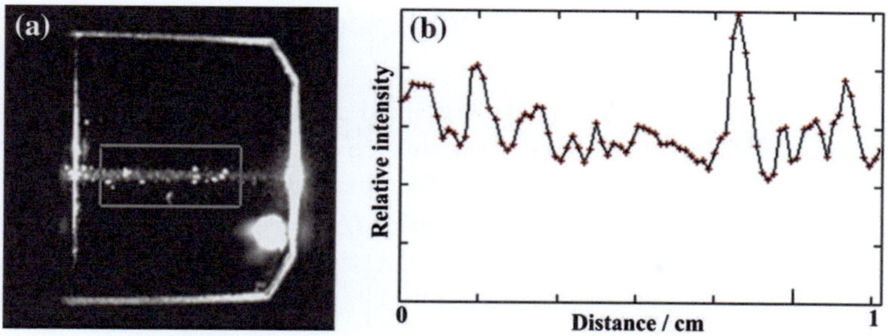

Fig. 4.15 **a** The photograph of one waveguide a polymer guiding layer and **b** the intensity of scattered light recorded along the propagation path

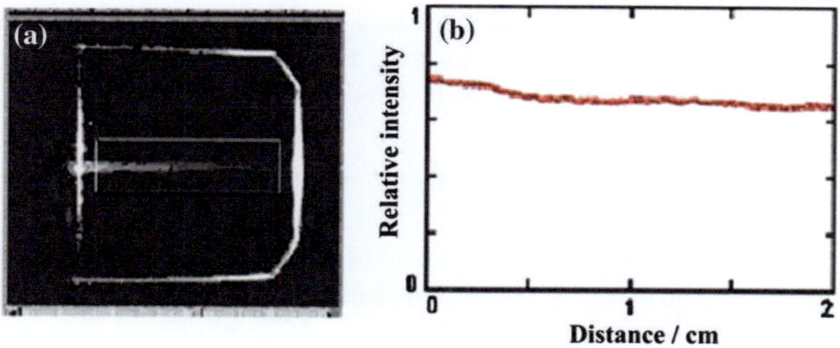

Fig. 4.16 The median filtering method resulted in **a** photograph of waveguide and **b** intensity of scattered light

The intensity of propagation light is described adequately by an exponential decay function dependent on the propagation distance

$$P = P_0 \exp(-\alpha z). \tag{4.70}$$

The attention coefficient α and loss ξ are calculated by the least square fitting method.

4.5 Evaluating Nonlinear Parameters of Waveguide

The study on the nonlinear waveguide optics has attracted considerable interest since it is capable of bounding the light energy in a small area and largely enhancing the density of optical power. In this subsection, we introduce the principle and experiment for evaluating the electro-optical (EO) and thermo-optical (TO) coefficients of a polymer layer by applying the prism-waveguide coupling system.

4.5.1 Measurement of the Electro-optic Coefficients

In the experiment, the incident angle θ is fixed at the middle point of the falling edge or the rising edge of one selected resonance dip (see Fig. 4.17). The variation of RI (Δn_{j1}) induced by the electric field (E) applied on the polymer layer will give rise to a change in the reflected light intensity (ΔR). For a prism-waveguide coupling system, the relation between ΔR and Δn_{j1} is [13]

$$\Delta R = k \cdot \frac{1}{n_3 \cos\theta} \cdot \Delta n_{j1}, \tag{4.71}$$

where n_3 is the RI of prism and k presents the slope of the linear area in the resonance dip.

Here, we give a general formula to evaluate the EO coefficients. For a polymer layer with both linear EO effect and quadratic EO effect, when an electric field (E) is applied perpendicular on it, and the field-induced RI change is

$$\Delta n_{j1} = -\frac{1}{2} n_1^3 \left(\gamma_{j1} E + s_{j1} E^2\right), \tag{4.72}$$

where γ_{j1} (contracted notation) is the element of linear EO coefficient tensor, and s_{j1} is the element of quadratic EO coefficient tensor with $j = 1$ or $j = 3$ for TE- or TM-guided mode, respectively. Substituting Eq. (4.72) into Eq. (4.71), we yield

$$\Delta R = -\frac{k n_1^3}{2 n_3 \cos\theta} \left(\gamma_{j1} E + s_{j1} E^2\right), \tag{4.73}$$

where n_1 is the RI of the polymer layer. The change of the reflected light intensity appears to follow a parabolic function of the applied electric field.

If the polymer layer has only the linear EO effect, its linear EO coefficient is derived as

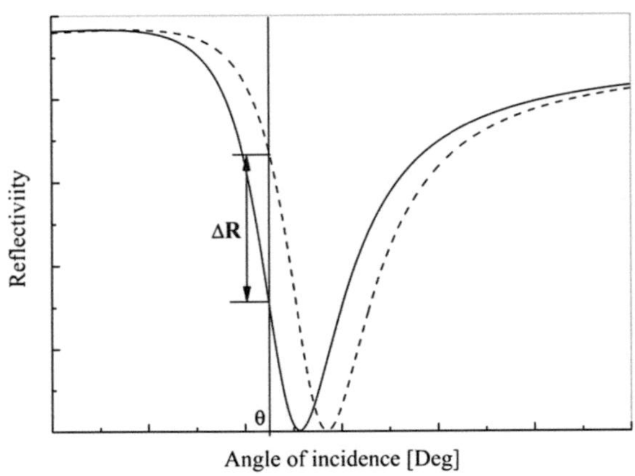

Fig. 4.17 The change of the reflected light intensity in response to the applied electric field

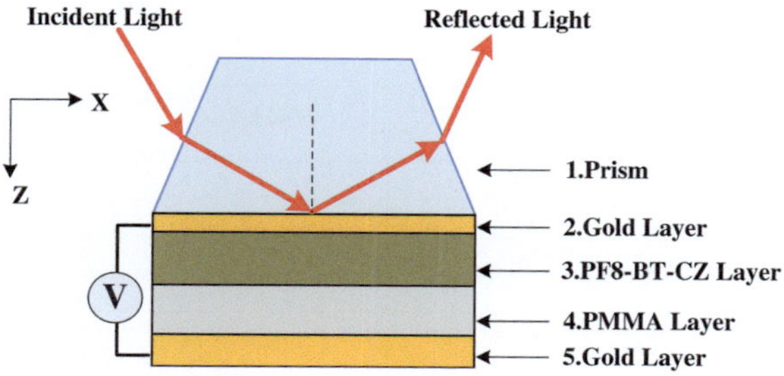

Fig. 4.18 Configuration for the simultaneous evaluation of the linear and quadratic EO coefficients

$$\gamma_{j1} = -\frac{2n_3 \cos\theta}{kn_1^3 E}\Delta R. \tag{4.74}$$

On the other hand, for an isotropic polymer layer, namely there is no linear EO effect and the quadratic EO is the first nonlinear effect, Eq. (4.73) is reduced to

$$\Delta R = -\frac{kn_1^3}{2n_3 \cos\theta} s_{j1} E^2. \tag{4.75}$$

It is clear that in this case the variation of the reflected intensity is quadratic to the applied electric field. The linear EO coefficient [13] and quadratic EO coefficient [14] can be separately measured because all parameters except γ_{j1} in Eq. (4.74) and s_{j1} in Eq. (4.75) are experimentally measurable.

A question arose here is how to evaluate the linear and quadratic EO coefficients simultaneously? The experimental setup [15] is shown in Fig. 4.18. It consists of (1) a prism; (2) a gold layer; (3) a polymeric guiding layer; (4) a buffer layer; and (5) a base gold layer. The applied triangular electric field has an amplitude ranging from $-E_m$ to E_m, the direction of which is perpendicular to the base surface of the polymer layer along the z direction. The electric field reaches its maximum value E_m in the $+z$ direction, where the variation of reflected intensity is proportional to $\gamma_{31}|E_m| + s_{31}|E_m|^2$, whereas the maximum value E_m in the $-z$ direction leads to variation of the reflected intensity which is proportional to $-\gamma_{31}|E_m| + s_{31}|E_m|^2$. Therefore, the variation of reflected intensity turns out as an unsymmetrical parabolic curve resulting from the combined action of linear EO and quadratic EO effects.

The polymeric guiding layer of the experimental sample is poly (9,9-dioctyl-2,7-fluorene-co-benzo[c] [1, 2, 5] thiadiazole-co-9-hexyl-3, 6-carbazole) (PF8-BT-CZ). In the experiment, a triangular AC electric field with 260 V peak-to-peak value of about 250 Hz was applied across the two gold electrodes of the sample. The oscilloscope traces of the applied electric field and the variation of reflected light intensity versus time are shown in Fig. 4.19. The applied

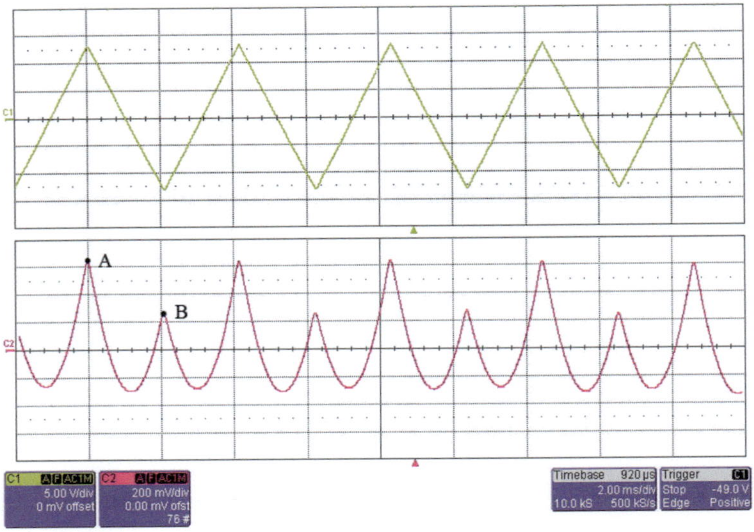

Fig. 4.19 Oscilloscope traces of applied voltage (*top*) and the reflected intensity versus time (*bottom*)

voltage is attenuated to be 10 % of its original value before put into the oscilloscope. With a linear increase or decrease in the applied electric field, the variation of reflected light intensity exhibited a parabolic curve with different maximal values at point A and point B because the variation of reflected light intensity is related not only to E^2 but also to E. By measuring the variation of reflected intensity at which the applied electric field reached its maximum E_m in the $+z$ and $-z$ directions, respectively, we obtained two groups of experimental data, which is substituted into Eq. (4.73) for constructing a set simultaneous equation to solve γ_{j1} and s_{j1}.

4.5.2 Evaluating the Thermo-Optical Coefficient of Polymer Layer

The steps of the conventional equipment Abbe refractometer to evaluate the TO coefficient of polymer layer are as follows: (1) measuring the RI of polymer layer under different temperatures and (2) fitting out the TO coefficient with the application of the obtained data. This method remains several issues to be improved. First, the polarization properties of TO coefficient cannot be studied since the separate measurement of TM polarization and TE polarization by the Abbe refractometer is impossible. Second, the Abbe refractometer is incapable of measuring the properties of polymer layer embedded in a waveguide.

The prism-waveguide coupling system to measure the TO coefficient of polymer layer [16, 17] is illustrated in Fig. 4.20. It consists of two components. The upper

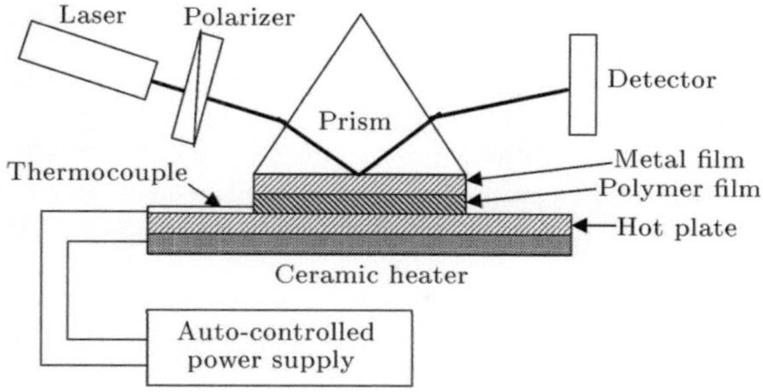

Fig. 4.20 Prism-waveguide coupling system to evaluate the TO coefficient

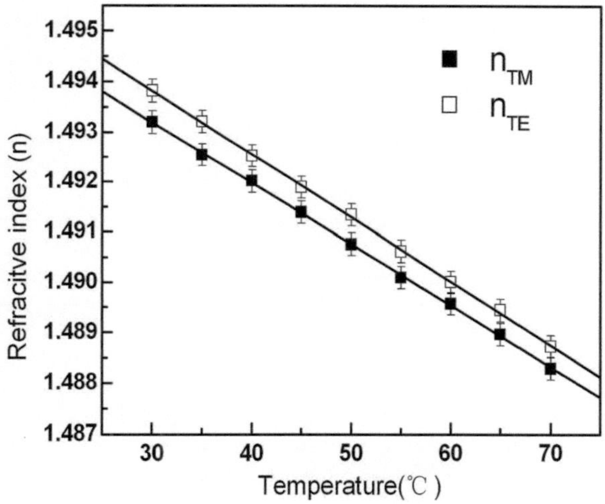

Fig. 4.21 The RI of PMMA layer under two linearly polarized states as functions of temperature

one is a glass prism with a thin metal film and a polymer layer coated on its base surface. The bottom one is an automatic temperature control unit. The two components are firmly placed together and mounted on a computer-controlled $\theta/2\theta$ goniometer. For the proof of concept, the RI of PMMA layer under two linearly polarized states as functions of the temperature is measured by the prism-waveguide coupling system, and the TO coefficients are calculated by the least squares fitting of the experimental data shown in Fig. 4.21. The result is well consistent with other measurement methods.

References

1. Z.Q. Cao, Q. Liu, Q.S. Shen, Simplified analysis for leaky modes in planar optical waveguides. Chin. Phys. Lett. **16**, 413 (1999)
2. H. Osterberg, L.W. Smith, Transmission of optical energy along surfaces: part II, inhomogeneous media. J. Opt. Soc. Am. **54**, 1078 (1964)

3. A. Otto, Excitation of non-radiative surface plasma waves in silver by the method of frustrated total reflection. Z. Phys. **216**, 398 (1968)
4. P.K. Tien, R. Ulrich, R.J. Martin, Modes of propagating light waves in thin deposited semiconductor films. Appl. Phys. Lett. **14**, 291 (1969)
5. X.B. Liu, Z.Q. Cao, P.F. Zhu, Q.S. Shen, X.G. Liu, Large positive and negative lateral optical beam shift in prism-waveguide coupling system. Phys. Rev. E **73**, 056617 (2006)
6. G. Borg, Eine Umkehrung der Strum-Liouvilleschen eigenwertaufgabe. Acta Math. **78**, 1 (1946)
7. J.M. White, P.F. Heidrich, Optical waveguide refractive index profiles determined from measurement of mode indices: a simple analysis. Appl. Opt. **15**, 151 (1976)
8. K.S. Chiang, Construction of refractive-index profiles of planar dielectric waveguides from the distribution of effective indexes. J. Lightwave Technol. **3**, 385 (1985)
9. Y. Ding, Z.Q. Cao, Q.S. Shen, Determination of optical waveguide refractive-index profiles with inverse analytic transfer matrix method. Opt. Quant. Electron. **36**, 489 (2004)
10. J.L. Merz, R.A. Logan, A.M. Sergent, Loss measurements in GaAs and AlxGa1-xAs dielectric waveguides between 1.1 ev and the energy gap. J. Appl. Phys. **47**, 1436 (1976)
11. H.P. Weber, F.A. Dunn, W. N. Leibolt, Loss measurements in thin-film optical waveguides. Appl. Opt. **12**, 755 (1973)
12. Y. Jiang, Z.Q. Cao, L.L. Qiu, Q.S. Shen, Y.L. Chen, Digital measurements and fabrication of low loss optical organic polymer waveguides. Acta Opt. Sinica **19**, 1142 (1999) (in Chinese)
13. Y. Jiang, Z Q. Cao, Q.S. Shen, X.M. Dou, Y.L. Chen, Y. Ozaki, Improved attenuated-total-reflection technique for measuring the electro-optic coefficients of nonlinear optical polymers. J. Opt. Soc. Am. B **17**, 805 (2000)
14. J.H. Zhou, X.X. Deng, Z.Q. Cao, Q.S. Shen, W. Wei, Z.J. Zhang, S.X. Xie, Determination of dc Kerr coefficients of polymer films with prism-optical waveguide configuration. Appl. Phys. Lett. **88**, 021106 (2006)
15. X.H. Zhu, X.X. Deng, H.G. Li, Z.Q. Cao, Q.S. Shen, W. Wei, F. Liu, Simultaneous evaluation of the linear and quadratic electro-optic coefficients of the nonlinear optical polymer by attenuated-total-reflection technique. J. Appl. Phys. **109**, 103105 (2011)
16. X. Li, Z.Q. Cao, Q.S. Shen, Y.F. Yang, Influence of dopant concentration on thermo-optic properties of PMMA composite. Mater. Lett. **60**, 1238 (2006)
17. X. Li, Z. Q. Cao, S.Q. Shen, Q.H. Meng, D.Y. Huang, K.P. Guo, L. Qiu, Y.Q. Shen, Anisotropy in thermo-optic coefficient of different polymer systems by attenuated total reflection configuration. Chin. Phys. Lett. **23**, 998 (2006)

Chapter 5
Surface Plasmon Wave

Abstract The pervious chapters introduce the dielectric slab waveguide, which is composed by ideal lossless materials; meanwhile, the widely used RF transmission lines and microwave waveguide is made of metallic lines, strip, and tubes. Metal is treated as a perfect conductor in the low-frequency range. Due to its collective electrons excitation, which is called as the plasmon, in the UV and visible region, metal can no longer be treated as a perfect conductor; it can still be applied to build low loss metal waveguide or metal–dielectric waveguide. In these two structures, the electromagnetic field takes the form of evanescent field. On the other hand, for noble metal such as gold, silver, and aluminum, their complex permittivity usually has a relatively larger real part than its imaginary part. In the near infrared and visible region, its real part is usually a large negative number

$$\varepsilon = \varepsilon_r + i\varepsilon_i,\ \varepsilon_r < 0,\ |\varepsilon_r| \gg \varepsilon_i. \tag{5.1}$$

Due to this optical property of metal, the surface plasmon wave (SPW) can propagate along its interface with dielectric, and the long-range SPW can be excited within a thin metal slab. These two surface waves illustrate different features from the conventional waveguide. Some unique characteristics can be valid, for example, the large range of the effective refractive index and the field enhancement effect. The propagation of SPW not only riches the traditional research field of the waveguide optics, but also finds a wide application in fields such as integrated optics, nonlinear optics, and molecular biology. This chapter first discusses the SPW bounded at a metallic interface and then analyzes the long-range SPW. The excitation method via attenuated total reflection method of these surface waves is introduced, and their applications are also discussed.

Keywords Surface plasmon wave · Long-range surface plasmon wave · Loss · Field enhancement effect

X. Wang et al., *Progress in Planar Optical Waveguides*,
Springer Tracts in Modern Physics 266, DOI 10.1007/978-3-662-48984-0_5

5.1 Optical Properties of Metal [1]

This section explains the origin of the complex permittivity of metal and interprets qualitatively the dispersion and absorption characteristics of metal in the visible frequencies by a simple physical model.

5.1.1 *The Permittivity Constant of Metal*

As is widely known, metal is a good conductor with a very high conductivity, and this property must be taken into consideration when dealing with metal's optical properties. Assuming the dielectric constant of a homogeneous isotropic and non-magnetic media to be $\varepsilon_0\varepsilon$, and its permeability to be μ_0, the conductivity to be σ, where ε denotes its relative permittivity, the Maxwell's equation can be recast into the following form

$$\left.\begin{array}{l}\nabla\times\vec{H}=\sigma\vec{E}+\varepsilon_0\varepsilon\frac{\partial\vec{E}}{\partial t}\\ \nabla\times\vec{E}=-\mu_0\frac{\partial\vec{H}}{\partial t}\\ \nabla\cdot\vec{E}=\dfrac{\rho}{\varepsilon_0\varepsilon}\\ \nabla\cdot\vec{H}=0\end{array}\right\}. \tag{5.2}$$

Taking the divergence on both sides of the first equation, and substituting the third equation into the first equation, there is

$$\frac{\partial\rho}{\partial t}+\frac{\sigma}{\varepsilon_0\varepsilon}\rho=0, \tag{5.3}$$

which follows

$$\rho=\rho_0\mathrm{e}^{-t/\tau}, \tag{5.4}$$

where

$$\tau=\frac{\varepsilon_0\varepsilon}{\sigma}, \tag{5.5}$$

is known as the relaxation time. It is clear that for any media with conductivity of σ, the free charge density ρ decays exponentially with time. Metal has a large conductivity and thus a small relaxation time, which is typically on the order of 10^{-18} s. This value is much smaller when compared with the photon with the same frequency, (about 10^{-15} s for the visible light). So it is reasonable to assume that the free charge density in metal is always equal to zero.

Taking the curl on both sides of Eq. (5.2), then inserting it into the first equation, and considering the fact that the charge density is always zero, one can get

$$\nabla^2 \vec{E} = \mu_0 \sigma \frac{\partial \vec{E}}{\partial t} + \varepsilon_0 \mu_0 \varepsilon \frac{\partial^2 \vec{E}}{\partial t^2}. \tag{5.6}$$

Assuming in all generality a harmonic time dependence $\vec{E} = \vec{E}_0 e^{-i\omega t}$ of angular frequency ω, the above equation can be simplified as

$$\nabla^2 \vec{E} + \widehat{k}^2 \vec{E} = 0, \tag{5.7}$$

with

$$\widehat{k}^2 = \frac{\omega^2}{c^2}\left(\varepsilon + i\frac{\sigma}{\omega\varepsilon_0}\right). \tag{5.8}$$

Note that only if we let the permittivity to take the form of a complex number

$$\widehat{\varepsilon} = \varepsilon + i\frac{\sigma}{\omega\varepsilon_0}, \tag{5.9}$$

then Eq. (5.7) will be mathematically equal with the Helmholtz equation. Combining Eqs. (5.7) and (5.8), it is obvious that the only difference between the plane wave propagating in a metallic media and in a transparent dielectric is that the usual real ε and k are replaced by complex $\widehat{\varepsilon}$ and $\widehat{k}$ in the former media.

5.1.2 Elementary Electronic Theory of Metal

Based on the elementary electronic theory of metal, a simple physical model can be established to qualitatively explain metal's dispersion and absorption in the visible regime of light and present some rough estimates of key parameters.

Usually, the bound electron dominates the dielectric characteristic and the free electron determines that of metal, except at ultraviolet and higher frequencies. At room temperature, a gas of free charges move randomly against a background of fixed positive ion lattice, and the metal is electrically neutral since the sum of negative free electrons per unit volume equals that of positive ions. Such a system of unbound positive and negative particles, whose overall charge is roughly zero, is called as plasma. When positive and negative charges separate, the ions can usually be treated as fixed due to its large mass, and the electrons oscillate harmonically in the electric field, and this phenomenon is called as plasma oscillations. The natural frequency is named as plasma frequency, which is denoted by ω_p.

Imagine a bulk of plasma, whose density of positive or negative charges is N, the electron gas shifts a distance x when subjected to an external electric field and the positive ions remains fixed. As can be seen in Fig. 5.1, thin layers of negative and opposite charge will appear on the left and right boundaries, respectively.

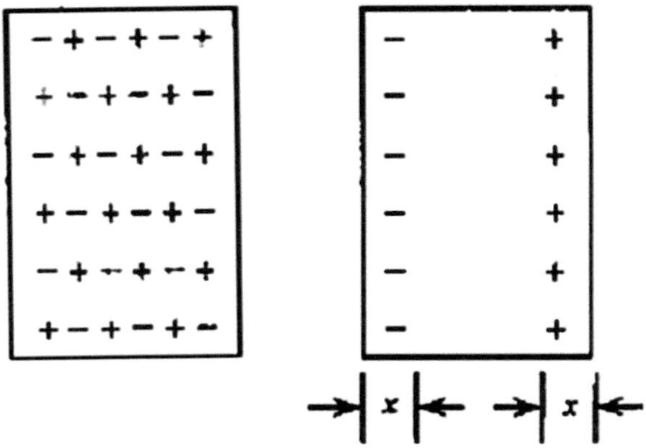

Fig. 5.1 Separation of positive and negative charges

The density of surface charges is Nex, where e represents the charge amount of a single electron. If an additional uniform electric field $E = -Nex/\varepsilon_0$ is exerted on the plasma and a restoring force will be loaded on a single electron,

$$F = -\frac{Ne^2}{\varepsilon_0}x. \tag{5.10}$$

As a result, the oscillation behavior of a free charge is rather similar with a spring oscillator with $k = Ne^2/\varepsilon_0$, and its natural angular frequency is

$$\omega_p = \sqrt{\frac{Ne^2}{\varepsilon_0 m}}. \tag{5.11}$$

For metal, ω_p is typically on the order of $10^{16}\,\mathrm{s}^{-1}$. When the electron gas in the metal is subjected to an external electric field with an angular frequency ω, it is expected that a forced vibration can be observed. So when we apply the plasma model, the motion of an electron will be governed by

$$m\ddot{r} + m\beta\dot{r} = eE, \tag{5.12}$$

with the term $m\beta\dot{r}$, which is proportional to its speed, denotes the damping suffered by the electron due to collisions and radiation.

Without any external electric field, the speed of an electron with an initial value of v_0 decays exponentially

$$v = \dot{r} = v_0 \cdot e^{-\beta\tau}, \tag{5.13}$$

where β is the attenuation constant and relates with the relaxation time by $\tau = 1/\beta$, which denotes the average time interval between two collisions of an electron, and it is typically of the order of 10^{-14} s.

In case of electrostatic field or in the low-frequency region, Eq. (5.12) can be rewritten as $m\beta\dot{r} = \mathrm{e}E$ by ignoring the electron's inertia. Assuming the free charge density is N, so the current density is

$$J = N\mathrm{e}\dot{r} = \frac{N\mathrm{e}^2}{m\beta}E, \tag{5.14}$$

and the corresponding conductivity is explicitly given by

$$\sigma = N\mathrm{e}^2/m\beta. \tag{5.15}$$

When applying a harmonic time-dependent external field $E = E_0\mathrm{e}^{-i\omega t}$, the particular solution of Eq. (5.12) takes the form $r = r_0\mathrm{e}^{-i\omega t}$ and can be solved as

$$r = -\frac{\mathrm{e}}{m(\omega^2 + i\beta\omega)} \cdot E,$$

the current density is

$$\vec{J} = N\mathrm{e}\frac{d\vec{r}}{dt} = \frac{N\mathrm{e}^2}{m(\beta - i\omega)}E, \tag{5.16}$$

and the frequency-dependent conductivity is

$$\sigma = \frac{Ne^2}{m(\beta - i\omega)}. \tag{5.17}$$

From the above equation, several conclusions can be drawn: (i) in case that $\omega \ll \beta$, σ can be approximated by $\sigma_0 = Ne^2/m\beta$, (ii) in case that $\omega \gg \beta$, i.e., the visible light region, σ has a larger imaginary part than its real part; (iii) in case of high frequency, σ becomes purely imaginary, and the permittivity $\hat{\varepsilon}$ is positive and real; consequently, metal acquires dielectric character.

As mentioned before, the free electrons govern the electromagnetic response of metal, so the first term ε on the right side of the complex permittivity Eq. (5.8) should be replaced by 1. Inserting the expression for σ and Eq. (5.11), one obtains

$$\hat{\varepsilon} = 1 - \frac{Ne^2}{\varepsilon_0 m\omega(\omega + i\beta)} = 1 - \frac{\omega_p^2}{\omega(\omega + i\beta)}. \tag{5.18}$$

In the visible part of the spectrum $\beta/\omega \ll 1$, so the real and imaginary components of the above complex permittivity are given by

$$\begin{cases} \mathrm{Re}\hat{\varepsilon} = 1 - \frac{\omega_p^2}{\omega^2 + \beta^2} \approx 1 - \left(\frac{\omega_p}{\omega}\right)^2, \\ \mathrm{Im}\hat{\varepsilon} = \frac{\omega_p^2 \beta}{\omega\left(\omega^2 + \beta^2\right)} \approx \frac{\beta}{\omega}\left(\frac{\omega_p}{\omega}\right)^2. \end{cases} \tag{5.19}$$

When $\omega < \omega_p$, the real component of $\hat{\varepsilon}$ is negative. Via inserting $\omega = 2.5 \times 10^{25}$ s^{-1}, $\beta = 10^{14}\,\mathrm{s}^{-1}$, $\omega_p = 10^{16}\,\mathrm{s}^{-1}$, so that $\omega_p/\omega = 4$, $\beta/\omega = 1/25$, and the real component of permittivity is negative $\mathrm{Re}\hat{\varepsilon} = -15$, which is 25 times of its imaginary part. This illustrates the fact that in the frequency region of the visible light, the real component is negative, and its magnitude is much bigger than that of the imaginary part. From Eq. (5.16), it is obvious that when $\omega \gg \beta$, there is a $\pi/2$ difference between the phases of the current density $\vec{J}$ and the electric field $\vec{E}$. When $\omega > \omega_p$, the real component of $\hat{\varepsilon}$ becomes real, and the bound electrons become more and more important when moving to higher frequency regime; therefore, the metal becomes more transparent and acts like a dielectric.

It should be pointed out that, the above model can only provide some qualitative explanation of the metal's optical response and present some approximate estimation of some key parameters, and a more strict theory will be needed to resort to quantum mechanics.

5.2 SPW on the Interface Between Metal and Dielectric

Due to the special optical property of metal, surface wave is different from the guided modes in an optical waveguide and can propagate along the interface between metal and dielectric. These surface waves have a wider range of its effective index and are evanescently confined in the perpendicular direction and therefore a strong field enhancement effect is obtained. They enrich the content of waveguide research, and thus, the metal-cladding waveguides find more application in fields of integrated optoelectronics, and biochemical detection, etc.

5.2.1 Excitation Condition of SPW

Let us consider an interface between two semi-infinite isotropic media, where the permittivity of the medium at $x > 0$ is written as $\varepsilon_1(\omega)$ and the permittivity of the medium at $x < 0$ is $\varepsilon_2(\omega)$ and set both media are non-magnetic, i.e., $\mu_1 = \mu_2 = \mu_0$. Based on the Maxwell's equations

$$\nabla \times \vec{E} = -\frac{\partial \vec{B}}{\partial t}, \tag{5.20}$$

$$\nabla \times \vec{H} = \frac{\partial \vec{D}}{\partial t}. \tag{5.21}$$

Via inserting Eq. (5.21) into Eq. (5.20), there is

$$\nabla \times \left(\nabla \times \vec{E}\right) + \mu_0 \frac{\partial^2 \vec{D}}{\partial t^2} = 0, \tag{5.22}$$

where $\vec{D} = \varepsilon_0 \varepsilon(\omega) \vec{E}$ and $\varepsilon(\omega)$ denotes the relative permittivity.

Since the surface waves are confined to the interface and decay in the perpendicular direction, the trial solution for the electric field should be

$$\left.\begin{array}{l} \vec{E}_1(\vec{r}, t) = \vec{E}_1^0 \exp(-\alpha_1 x) \exp[i(\beta z - \omega t)]\ (x > 0) \\ \vec{E}_2(\vec{r}, t) = \vec{E}_2^0 \exp(\alpha_2 x) \exp[i(\beta z - \omega t)]\ (x < 0) \end{array}\right\}. \tag{5.23}$$

Inserting Eq. (5.23) into Eq. (5.22) and combing the divergence of the electric field

$$\nabla \cdot \vec{E} = 0,$$

one will have

$$\left.\begin{array}{l} \vec{E}_1(x) = \left(\frac{i\beta}{\alpha_1} E_{1z}^0, E_{1y}^0, E_{1z}^0\right) \exp(-\alpha_1 x)\ (x > 0) \\ \vec{E}_2(x) = \left(-\frac{i\beta}{\alpha_2} E_{2z}^0, E_{2y}^0, E_{2z}^0\right) \exp(\alpha_2 x)\ (x < 0) \end{array}\right\}. \tag{5.24}$$

In the above equations, the propagation factor $\exp[i(\beta z - \omega t)]$ has been omitted for simplicity. The attenuation coefficients α_1 and α_2 are given by

$$\left.\begin{array}{l} \alpha_1^2 = \beta^2 - k_0^2 \varepsilon_1 \\ \alpha_2^2 = \beta^2 - k_0^2 \varepsilon_2 \end{array}\right\}. \tag{5.25}$$

Similarly, the expressions for the magnetic field are

$$\left.\begin{array}{l} \vec{H}_1(x) = \frac{i}{\omega \mu_0} \left(i\beta E_{1y}^0, \frac{k_0^2 \varepsilon_1}{\alpha_1} E_{1z}^0, \alpha_1 E_{1y}^0\right) \exp(-\alpha_1 x)\ (x > 0) \\ \vec{H}_2(x) = \frac{i}{\omega \mu_0} \left(i\beta E_{2y}^0, -\frac{k_0^2 \varepsilon_2}{\alpha_2} E_{2z}^0, -\alpha_2 E_{2y}^0\right) \exp(\alpha_2 x)\ (x < 0) \end{array}\right\}. \tag{5.26}$$

Applying the boundary conditions, if follows that

$$E_{1y}^0 = E_{2y}^0, \tag{5.27}$$

$$E_{1z}^0 = E_{2z}^0, \tag{5.28}$$

$$\frac{\varepsilon_1}{\alpha_1} E_{1z}^0 = -\frac{\varepsilon_2}{\alpha_2} E_{2z}^0, \tag{5.29}$$

$$\alpha_1 E_{1y}^0 = -\alpha_2 E_{2y}^0. \tag{5.30}$$

Three conclusions can be drawn based on the above equations

1. Since both α_1 and α_2 are real positive number, combing Eqs. (5.27) and (5.30), it is clear that

$$E_{1y}^0 = E_{2y}^0 = 0, \tag{5.31}$$

which shows that the SPWs are always TM polarized.

2. Combing Eqs. (5.28) and (5.29), the dispersion expression of the SPW takes the form

$$\frac{\alpha_1}{\alpha_2} = -\frac{\varepsilon_1}{\varepsilon_2}, \tag{5.32}$$

which demonstrates that this kind of waves can only be excited when the permittivity of the two media take opposite signs.

3. Inserting Eq. (5.25) into Eq. (5.32), the effective index of the SPW can be written as

$$\frac{\beta}{k_0} = \sqrt{\frac{\varepsilon_1 \varepsilon_2}{\varepsilon_1 + \varepsilon_2}}. \tag{5.33}$$

Based on the second conclusion, if one of the media in Fig. 5.2 is metal with negative permittivity (strictly speaking, the metal's permittivity is complex, but its

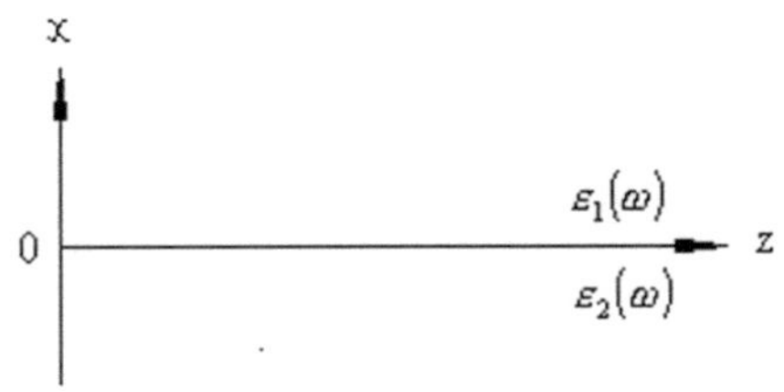

Fig. 5.2 Interface between two media of different permittivities

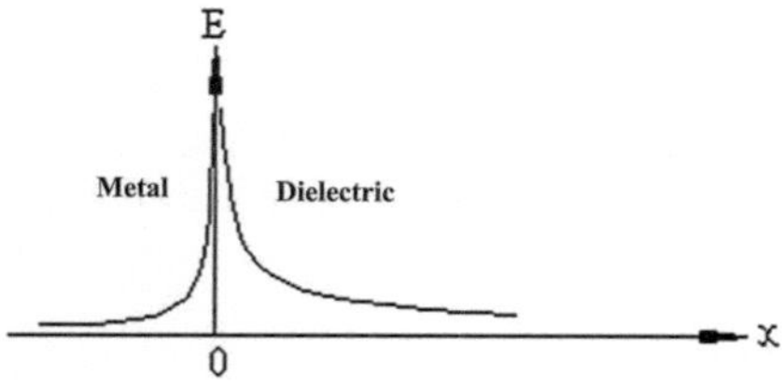

Fig. 5.3 Electric field distribution of the SPW

imaginary component has a much smaller magnitude), the excitation condition of the SPW can be fulfilled. And the electromagnetic field distribution along the x-axis perpendicular to the interface is plotted in Fig. 5.3.

5.2.2 *Loss*

The permittivity of metal is complex, where its imaginary part denotes the absorption of the light energy inside the metal. Without losing generality, let us set the $\varepsilon_2(\omega)$ in Fig. 5.2 to be complex, there is

$$\varepsilon_2 = \varepsilon_{r2} + i\varepsilon_{i2}, \ (\varepsilon_{i2} > 0). \tag{5.34}$$

The introduction of the complex permittivity of metal results in a complex wavenumber for the SPW, while its imaginary component represents the loss suffered by the surface during propagation. Let us consider two simple cases:

1. The first case is $\varepsilon_{i2} = 0$, which represents that the metal is non-absorptive. (This situation is physically unrealistic, but can still be served as an approximation when $|\varepsilon_{r2}| \gg \varepsilon_{i2}$ holds.) In this case, there is $\varepsilon_2 = \varepsilon_{r2}$, and in general we can also assume that $|\varepsilon_{r2}| > \varepsilon_1$, consequently the propagation constant of the SPW is

$$\beta_r = k_0\sqrt{\frac{\varepsilon_1\varepsilon_{r2}}{\varepsilon_1 + \varepsilon_{r2}}}, \tag{5.35}$$

$$\beta_i = 0. \tag{5.36}$$

Since the propagation constant is purely real, the corresponding attenuation coefficients α_1 and α_2 are also real, which read now

$$\alpha_1 = k_0\varepsilon_1\sqrt{\frac{1}{|\varepsilon_{r2}| - \varepsilon_1}}, \tag{5.37}$$

$$\alpha_2 = k_0|\varepsilon_{r2}|\sqrt{\frac{1}{|\varepsilon_{r2}| - \varepsilon_1}}. \tag{5.38}$$

2. For the second case $\varepsilon_{i2} \neq 0$, then all the quantities β, α_1, and α_2 are complex, so based on Eq. (5.33) they can be solved analytically and be written as

$$\beta_r = k_0\left[\frac{\varepsilon_1}{(\varepsilon_1+\varepsilon_{r2})^2+\varepsilon_{i2}^2}\right]^{1/2} \cdot \left[\frac{\varepsilon_e^2+\left(\varepsilon_e^4+\varepsilon_1^2\varepsilon_{i2}^2\right)^{1/2}}{2}\right]^{1/2}, \tag{5.39}$$

$$\beta_i = k_0\left[\frac{\varepsilon_1}{(\varepsilon_1+\varepsilon_{r2})^2+\varepsilon_{i2}^2}\right]^{1/2} \cdot \frac{\sqrt{2}\varepsilon_{i2}\varepsilon_1}{2\left[\varepsilon_e^2+\left(\varepsilon_e^4+\varepsilon_1^2\varepsilon_{i2}^2\right)^{1/2}\right]^{1/2}}, \tag{5.40}$$

where

$$\varepsilon_e^2 = \varepsilon_{r2}^2 + \varepsilon_{i2}^2 + \varepsilon_1\varepsilon_{r2}. \tag{5.41}$$

Usually, the permittivity of metal satisfies $\varepsilon_{r2} < 0$, $|\varepsilon_{r2}| \gg \varepsilon_{i2}$, and $|\varepsilon_{r2}| \gg \varepsilon_1$, which can further simplify the above two formula into the following form

$$\beta_r \approx k_0\sqrt{\frac{\varepsilon_1\varepsilon_{r2}}{\varepsilon_1+\varepsilon_{r2}}}, \tag{5.42}$$

$$\beta_i \approx k_0\frac{\varepsilon_{i2}\varepsilon_1}{2\varepsilon_{r2}(\varepsilon_1+\varepsilon_{r2})} \cdot \sqrt{\frac{\varepsilon_1\varepsilon_{r2}}{\varepsilon_1+\varepsilon_{r2}}}. \tag{5.43}$$

In view of Eqs. (5.42) and (5.43), we can see that $\beta_r > k_0\sqrt{\varepsilon_1}$ and β_i is proportional to ε_{i2}. Comparing with Eq. (5.35), it is clear that if we ignore the imaginary part of metal's permittivity, the characteristics of the SPW will not be varied except that its propagation loss is omitted.

5.2.3 The Excitation Scheme of the SPW

From the previous sections, we know there are certain conditions for the excitation of SPW at the interface between metal and dielectric. The pioneer work was done by Otto et al. [2] and Kretschmann et al. [3] on the issue of optical excitation of the SPW. They adopted the prism coupling structure as shown in Fig. 5.4, where the refractive index of prism is sufficiently large. The incident angle at the bottom of the prism can be tuned, so that the wave vector along the interface of the incident TM wave can be matched with the wave vector of the SPW. By adjusting the thickness

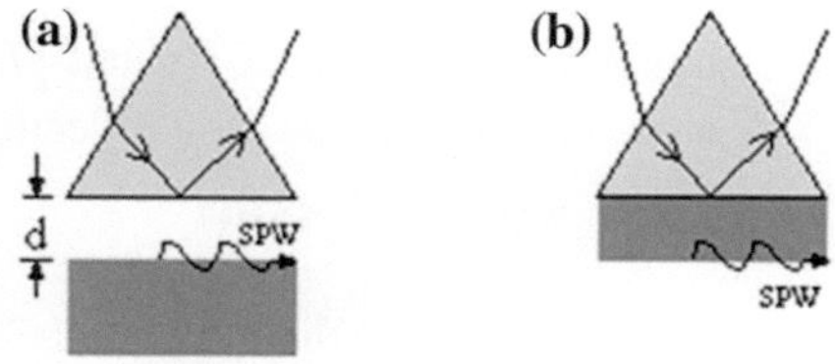

Fig. 5.4 The prism coupling method for SPW

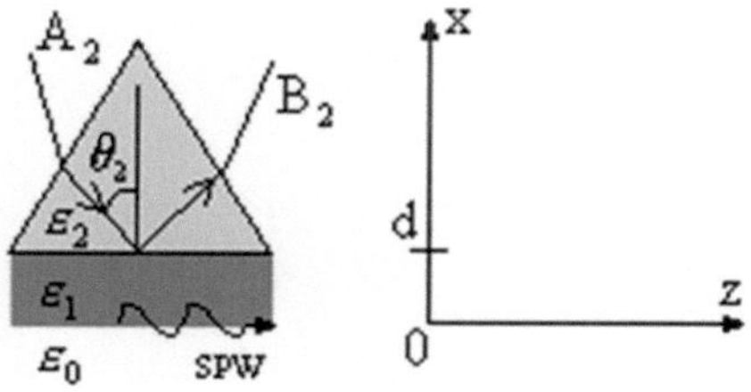

Fig. 5.5 Actual ATR structure

of the air gap in Otto's setup or the thickness of the metal film in the Kretschmann's configuration, the SPW can be effectively excited, namely, most of the incident power is coupling into the surface waves and the intensity of the reflected light beam is significantly reduced. As a result, a resonance absorption dip can be observed in the reflection spectrum. This excitation method is known as the attenuated total reflection technology, or ATR technology for short.

Let us take the Kretschmann's structure, for example, to derive the expression of its reflection probability and consider the actual structure and the coordinate as plotted in Fig. 5.5. The permittivity of the prism and dielectric is set to be ε_2 and ε_0, respectively, and the complex permittivity of metal to be $\varepsilon_1 = \varepsilon_{r1} + i\varepsilon_{i1}$, and the thickness of the metal film to be *d*. The SPW propagates along the *z*-axis, and the incident angle at the prism/metal interface is θ_2. According to the SPW's characteristics, the magnetic field distribution can be written as

$$Hy(x) = \begin{cases} A_2 e^{\alpha_2 (x-d)} + B_2 e^{-\alpha_2 (x-d)}, & (x > d), \\ A_1 e^{\alpha_1 x} + B_2 e^{-\alpha_1 x}, & (0<x<d), \\ A_0 e^{\alpha_0 x}, & (x<0). \end{cases} \tag{5.44}$$

with

$$\begin{aligned} \alpha_j &= \left(\beta^2 - k_0^2 \varepsilon_j\right)^{1/2}, \ (j = 0, 1, 2), \\ \beta &= k_0 \sqrt{\varepsilon_2} \sin\theta_2. \end{aligned} \tag{5.45}$$

The optical field in the prism is oscillating, the parameter α_2 should be imaginary; meanwhile, the parameter α_1 corresponding to metal is complex due to the complex permittivity ε_1. Using the continuous condition of Hy and $\frac{1}{\varepsilon}Hy'$ at the interface of $x = 0$ and $x = d$ for TM mode, the expression for reflection coefficient can be solved as

$$R = \left|\frac{B_2}{A_2}\right|^2 = \left|\frac{\gamma_{12} + \gamma_{01}e^{-2\alpha_1 d}}{1 + \gamma_{12}\gamma_{01}e^{-2\alpha_1 d}}\right|^2. \tag{5.46}$$

with

$$\gamma_{01} = \frac{\varepsilon_0\alpha_1 - \varepsilon_1\alpha_0}{\varepsilon_0\alpha_1 + \varepsilon_1\alpha_0}, \tag{5.47}$$

$$\gamma_{12} = \frac{\varepsilon_1\alpha_2 - \varepsilon_2\alpha_1}{\varepsilon_1\alpha_2 + \varepsilon_2\alpha_1}. \tag{5.48}$$

After the specific parameters of the experimental setup are fixed, Eq. (5.46) can be applied to plot the reflection probability R as a function of incident angle θ_2. A concrete example is presented in Fig. 5.6 with parameters $\varepsilon_0 = 1.0$, $\varepsilon_1 = -18.0 + i0.7$, $\varepsilon_2 = 3.24$, $d = 50.0$ nm, and $\lambda = 632.8$ nm. When the incident angle θ_2 is equal to a specific angle of θ_{ATR}, the reflection probability R reaches its minimum. This denotes that the z-component of the incident wave vector equals the propagation constant β_{ATR} of the SPW, the surface wave is excited, and energy is transferred to the metal/dielectric interface. The half width of the resonance represents the energy loss induced by the metal absorption.

The permittivity and the thickness of the metal film are the key parameters in many related experiments, so it is usually required that these parameters must be measured. The conventional measurement method is the double-wavelength excitation method [4], i.e., using two wavelength to excite SPW and measuring their reflection spectrum, then determining the complex permittivity and thickness via

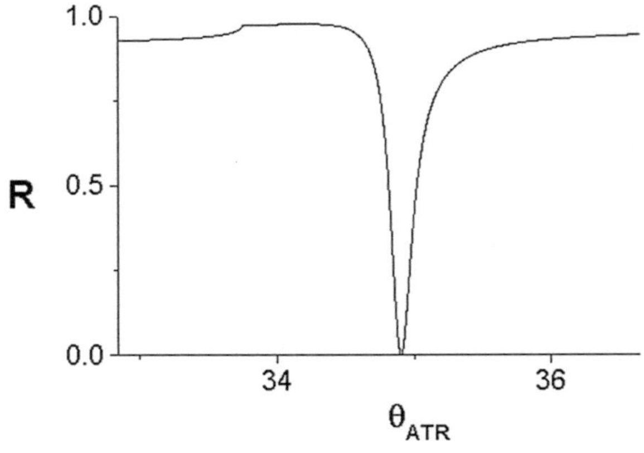

Fig. 5.6 Reflection spectrum of ATR

numerically fitting method. Similarly, we proposed another method to test these quantities via changing the dielectric adjacent the metal [5] and yield good results. Note that SPW can also be excited via grating coupling method, which facilitates the miniaturization of the related optoelectronic devices.

5.2.4 Field Enhancement Effect

As mentioned before, the SPW is a kind of electromagnetic wave propagates along the metal/dielectric interface and decays exponentially from the interface. In short, the field intensity is strongest at the interface. Taking Kretschmann's structure, for example, setting the refractive index of the prism to be $n_2 = 1.515$, and the permittivity of the 45-nm-thick silver film to be $\varepsilon_1 = -17.3 + i0.68$, and the refractive index for air $n_0 = 1.0$, the electric field distribution of the SPW is plotted in Fig. 5.7. If we normalize the electric amplitude of the incident light in the prism, at the boundary $|E_x|$ is enhanced by 18.8 times, while $|E_z|$ is enhanced by 4.2 times and the enhancement of the Poynting vector is as high as 335.4 times. Due to this field enhancement effect, the SPW obtains a wide range of applications in fields of biology, chemistry, materials science, and designing optoelectronic devices.

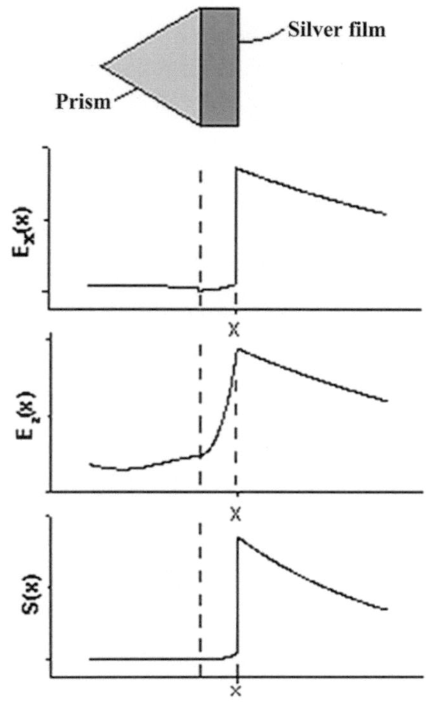

Fig. 5.7 Field enhancement effect in the Kretschmann's structure

5.3 Measurement of Metal Film's Thickness and Permittivity by Double-Wavelength Method

The optical characteristics of metal are dominated by its frequency-dependent complex permittivity $\varepsilon(\omega)$, and various methods have been developed to determine $\varepsilon(\omega)$ experimentally. Among all these methods, the SPW technology is particularly suitable to study the metal film with nanometer-scale thickness. During the late seventies of the last century, Lopez-Rios and Vuye [6] proposed a numerical fitting method to determine the metal film thickness d and permittivity $\varepsilon(\omega)$ simultaneously via exciting the SPW. But this method requires that the functional form of $\varepsilon(\omega)$ should be given in advance, but that it is unrealistic in most cases. So there is a need to measure these key parameters of a metal film experimentally without the knowledge of the functional form of $\varepsilon(\omega)$. In this section, a widely applied double-wavelength method is introduced [4].

5.3.1 *Measurement Principle*

Let us consider the excitation of SPW via the Kretschmann's configuration as plotted in Fig. 5.8 and denote the permittivity of air, metal film and prism as ε_0, ε_1 and ε_2, respectively, the thickness of metal film as d. A laser of TM polarization and angular frequency ω is incident onto the prism/metal interface via incident angle θ. When the wave vector of incident matches β^L via adjusting incident angle, the SPW can be excited, the energy can be transferred into the surface wave and a resonance dip plotted in Fig. 5.9 can be formed in the reflection spectrum. θ_{ATR} represents the resonance angle, ω_θ is the width of the resonance dip, and $R_{\min}$ denotes the minimum of the reflection probability.

The wave vector of the SPW is complex due to the metal's absorption and the energy leakage from the prism. On the other hand, the wave vector of the incident light along the interface is purely real, so the match between these two wave vectors is approximate, where

$$\beta = \mathrm{Re}\left(\beta^L\right). \tag{5.49}$$

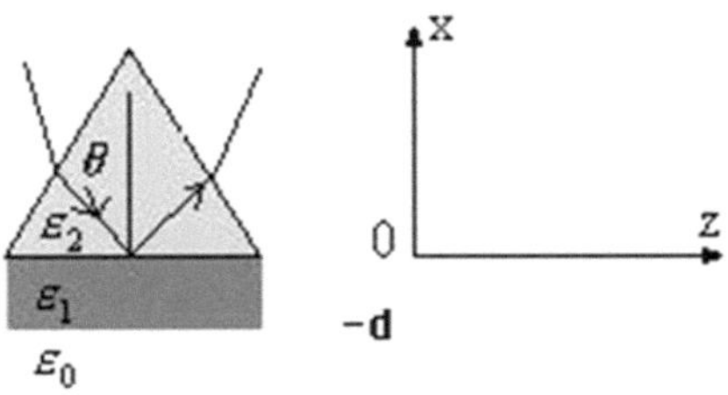

Fig. 5.8 The excitation of SPW via Kretschmann's structure

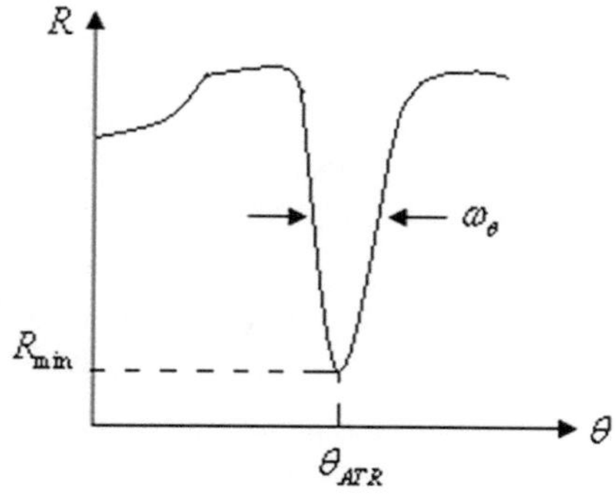

Fig. 5.9 ATR spectrum of a SPW

As a result, the resonance angle is

$$\theta_{\mathrm{ATR}} = \sin^{-1}\left[\frac{\mathrm{Re}\left(\beta^L\right)}{k_0\sqrt{\varepsilon_2}}\right]. \tag{5.50}$$

The derivation of the reflection probability is not difficult, and R can be written as

$$R = \left|\frac{r_{12} + r_{10}\exp(-2\alpha_1 d)}{1 + r_{10}r_{12}\exp(-2\alpha_1 d)}\right|^2, \tag{5.51}$$

where

$$r_{10} = \frac{\varepsilon_1\alpha_0 - \varepsilon_0\alpha_1}{\varepsilon_1\alpha_0 + \varepsilon_0\alpha_1}, \tag{5.52}$$

$$r_{21} = \frac{\varepsilon_1\alpha_2 - \varepsilon_2\alpha_1}{\varepsilon_1\alpha_2 + \varepsilon_2\alpha_1}, \tag{5.53}$$

and

$$\begin{cases} \alpha_0 = \left(\beta^2 - k_0^2\varepsilon_0\right)^{1/2}, \\ \alpha_1 = \left(\beta^2 - k_0^2\varepsilon_1\right)^{1/2}, \\ \kappa_2 = \left(k_0^2\varepsilon_2 - \beta^2\right)^{1/2}. \end{cases} \tag{5.54}$$

As the incident angle θ approaches the resonance angle θ_{ATR}, a good approximation of the reflection probability R can be derived as

$$R = 1 - \frac{4\mathrm{Im}\left(\beta^0\right)\mathrm{Im}\left(\Delta\beta^L\right)}{\left[\beta - \mathrm{Re}\left(\beta^L\right)\right]^2 + \left[\mathrm{Im}\left(\beta^L\right)\right]^2}, \tag{5.55}$$

where

$$\beta^L = \beta^0 + \Delta\beta^L. \tag{5.56}$$

In the above formula, β^L is the complex wave vector with the presence of the prism, while β^0 is the wave vector of the free SPW without the prism. So it is clear that $\Delta\beta^L$ is the perturbation on β^0 due to the energy leakage from the prism. For the metal film, there is $\varepsilon_1 = \varepsilon_{1r} + \varepsilon_{1i}$ and $|\varepsilon_{1r}| \gg \varepsilon_{1i}$, which follows

$$\beta^0 = k_0\sqrt{\frac{\varepsilon_1\varepsilon_0}{\varepsilon_1+\varepsilon_0}} \approx k_0\sqrt{\frac{\varepsilon_{1r}\varepsilon_0}{\varepsilon_{1r}+\varepsilon_0}} + ik_0\sqrt{\frac{\varepsilon_{1r}\varepsilon_0}{\varepsilon_{1r}+\varepsilon_0}} \cdot \frac{\varepsilon_{1i}\varepsilon_0}{2\varepsilon_{1r}(\varepsilon_{1r}+\varepsilon_0)}. \tag{5.57}$$

$$\Delta\beta^L = k_0(r_{21})_{\beta=\beta^0}\frac{2}{\varepsilon_0-\varepsilon_1}\left(\frac{\varepsilon_0\varepsilon_1}{\varepsilon_0+\varepsilon_1}\right)^{3/2}\exp\left[-2k_0 d\frac{\varepsilon_1}{(\varepsilon_1+\varepsilon_0)^{1/2}}\right]. \tag{5.58}$$

The imaginary components of β^0 and $\Delta\beta^L$ are called as the intrinsic and the radiation attenuation, respectively. The former is the Joule loss in the metal, while the latter represents the energy leakage from the prism. These two kinds of loss determine the half width of the absorption dip in the reflection spectrum, and we can write

$$\mathrm{Im}(\beta^L) = \Delta(k_0\sqrt{\varepsilon_2}\sin\theta_{\mathrm{ATR}}) = k_0\sqrt{\varepsilon_2}\cos\theta_{\mathrm{ATR}} \cdot \Delta\theta, \tag{5.59}$$

and

$$\omega_\theta = 2\Delta\theta = \frac{2\mathrm{Im}(\beta^L)}{k_0\sqrt{\varepsilon_2}\cos\theta_{\mathrm{ATR}}}. \tag{5.60}$$

Equation (5.49) holds at the resonance angle θ_{ATR}, and the reflectivity reaches its minimum value

$$R_{\min} = 1 - \frac{4\eta}{(1+\eta)^2}, \tag{5.61}$$

where

$$\eta = \mathrm{Im}(\beta^0)/\mathrm{Im}(\Delta\beta^L). \tag{5.62}$$

Based on the above analysis and the experimentally measured ATR spectrum, the complex permittivity ε_1 and thickness d of the metal film can be determined.

5.3.2 Experiment and Measurement

The experiment process includes the preparation of the sample and the measurement of the ATR spectrum. The detailed steps of the calculation of the complex ε_1 and d are listed below:

1. Inserting the measured θ_{ATR} into Eq. (5.49), we can solve the $\mathrm{Re}(\beta^L)$;
2. Since $\mathrm{Re}(\Delta\beta^L) \ll \mathrm{Re}(\beta^0)$, we can assume that $\mathrm{Re}(\beta^0) = \mathrm{Re}(\beta^L)$;
3. Determine ε_{1r} via the real part of Eq. (5.57);
4. Inserting the measured θ_{ATR} and W_θ into Eq. (5.60) to determine the term $\mathrm{Im}(\beta^L)$. Note that $\mathrm{Im}(\beta^L) = \mathrm{Im}(\beta^0) + \mathrm{Im}(\Delta\beta^L)$;
5. Substituting the measured $R_{\min}$ into Eq. (5.61) to solve the value of $\mathrm{Im}(\beta^0)/\mathrm{Im}(\Delta\beta^L)$;
6. The terms $\mathrm{Im}(\beta^0)$ and $\mathrm{Im}(\Delta\beta^L)$ can be solved based on the two steps above;
7. Combine the value of $\mathrm{Im}(\beta^0)$, ε_{1r}, and Eq. (5.57), ε_{1i} is available;
8. Finally, based on $\mathrm{Im}(\Delta\beta^L)$, θ_{ATR}, ε_{1r}, and ε_{1i}, the thickness d can be obtained using Eq. (5.58).

Since Eq. (5.61) has two reciprocal sets of solution, where one set corresponds to the case that intrinsic attenuation is greater than the radiation attenuation (less coupling situation); the other set presents the contrary case, i.e., the radiation attenuation is greater than the intrinsic attenuation (over coupling). The thickness corresponding to the former set of solution is larger than the other one. In the first-order approximation, these two sets of solutions are difficult to discern based on the ATR spectrum, since both sets can be used as initial values to produce very good fitting curves. In order to determine a set of real solutions, two different ATR curves excited by two different wavelength λ_1 and λ_2 must be measured. Although each wavelength can produce two sets of ε_1 and d, the thickness of the metal film is wavelength independent, we can always find two values of d corresponding to different wavelengths are closed to each other. Consequently, the exact thickness and the metal permittivity corresponding to the two wavelength λ_1 and λ_2 can be obtained from these two sets of solutions.

For example, if the two sets of solutions corresponding to wavelength $\lambda_1 = 632.8$ nm are

$$1.\ \varepsilon_1 = -17.45 + i0.92, \quad d = 38.7\ \text{nm}$$
$$2.\ \varepsilon_1 = -16.72 + i1.66, \quad d = 48.3\ \text{nm}$$

and for $\lambda_2 = 488.0$ nm, the two sets of solutions are

$$1.\ \varepsilon_1 = -8.39 + i0.60, \quad d = 56.7\ \text{nm}$$
$$2.\ \varepsilon_1 = -8.31 + i0.68, \quad d = 47.1\ \text{nm}$$

The values of thickness d in the second sets corresponding to both wavelength λ_1 and λ_2 are close to each other. So we can determine that the properties of the metal film at the two frequencies are

$$\varepsilon_1 = -16.72 + i1.66, \quad d = 47.7\,\text{nm}, \quad \lambda = 632.8\,\text{nm}$$

$$\varepsilon_1 = -8.31 + i0.68, \quad d = 47.7\,\text{nm}, \quad \lambda = 488.0\ \text{nm}$$

5.4 Long-Range SPW of a Metal Film Structure

Different from the SPW, the long-range SPW (LRSPW) can be excited in a symmetric structure. The confinement of the coupled SPW to the metal film decreases, and most energy is stored in the homogeneous dielectric. Consequently, a drastically increase in the propagation length is observed, and this kind of surface wave is called as the long-range SPW.

5.4.1 Dispersion

Consider the metal film structure as shown in Fig. 5.10, regions 1 and 3 denote non-absorptive media with permittivity written as ε_1 and ε_3, respectively. Region 2 is the metal film of thickness d and complex permittivity $\varepsilon_2 = \varepsilon_{r2} + i\varepsilon_{i2}$. The optical characteristics of metal implies that $\varepsilon_{r2} < 0$, $|\varepsilon_{r2}| \gg \varepsilon_{i2}$ and $|\varepsilon_{r2}| > \varepsilon_1,\ \varepsilon_3$. For the convenience of analysis, let us ignore the absorption at beginning, so there is $\varepsilon_2 = \varepsilon_{r2}$ and $\varepsilon_{i2} = 0$. We will discuss two different situations: (1) asymmetric structure and (2) symmetric structure.

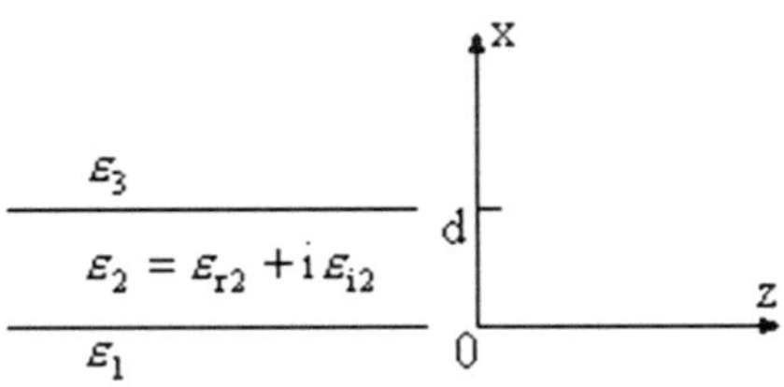

Fig. 5.10 Metal film structure

1. Asymmetric structure

Consider the TM wave travels along the z direction at the thin metal film, where the field distribution in the three media is given by

$$Hy(x) = \begin{cases} A_1 e^{\alpha_1^0 x}, & x<0, \\ A_2 e^{\alpha_2^0 x} + B_2 e^{-\alpha_2^0 x}, & 0<x<d, \\ B_3 e^{-\alpha_3^0 (x-d)}, & x>d. \end{cases} \tag{5.63}$$

The superscript (0) denotes the ideal system without absorption, and there are

$$\left.\begin{aligned} \alpha_1^0 &= \left(\beta^2 - k_0^2 \varepsilon_1\right)^{1/2} \\ \alpha_2^0 &= \left(\beta^2 - k_0^2 \varepsilon_2\right)^{1/2} \\ \alpha_3^0 &= \left(\beta^2 - k_0^2 \varepsilon_3\right)^{1/2} \end{aligned}\right\}. \tag{5.64}$$

Based on the transfer matrix method, the matrix equation for the magnetic field is

$$\begin{bmatrix} H_y(0) \\ \frac{1}{\varepsilon_1} H_y'(0) \end{bmatrix} = M_2 \begin{bmatrix} H_y(d) \\ \frac{1}{\varepsilon_3} H_y'(d) \end{bmatrix}, \tag{5.65}$$

where

$$M_2 = \begin{bmatrix} \cosh\left(\alpha_2^0 d\right) & -\frac{\varepsilon_2}{\alpha_2^0} \sinh\left(\alpha_2^0 d\right) \\ -\frac{\alpha_2^0}{\varepsilon_2} \sinh\left(\alpha_2^0 d\right) & \cosh\left(\alpha_2^0 d\right) \end{bmatrix}, \tag{5.66}$$

Using Eq. (5.63) and inserting the boundary conditions into Eq. (5.65), the matrix equation reduces to

$$\begin{pmatrix} -\frac{\alpha_1^0}{\varepsilon_1} & 1 \end{pmatrix} M_2 \begin{pmatrix} 1 \\ -\frac{\alpha_3^0}{\varepsilon_3} \end{pmatrix} = 0. \tag{5.67}$$

The dispersion relation of the LRSPW on the asymmetric thin metal film structure can be obtained by solving the above equation, namely

$$\tanh\left(\alpha_2^0 d\right) = -\frac{\frac{\varepsilon_2 \alpha_1^0}{\varepsilon_1 \alpha_2^0} + \frac{\varepsilon_2 \alpha_3^0}{\varepsilon_3 \alpha_2^0}}{1 + \frac{\varepsilon_2 \alpha_1^0}{\varepsilon_1 \alpha_2^0} \cdot \frac{\varepsilon_2 \alpha_3^0}{\varepsilon_3 \alpha_2^0}}. \tag{5.68}$$

2. Symmetric structure

If the regions 1 and 3 contain the same materials, i.e., $\varepsilon_1 = \varepsilon_3$, then Eq. (5.68) can be simplified as

$$\tanh\left(\alpha_2^0 d\right) = -\frac{2\dfrac{\varepsilon_2\alpha_1^0}{\varepsilon_1\alpha_2^0}}{1+\left(\dfrac{\varepsilon_2\alpha_1^0}{\varepsilon_1\alpha_2^0}\right)^2}. \tag{5.69}$$

In this case, Eq. (5.69) can be split into two equations: one corresponds to the even mode, which fulfills the inequalities $\left|\frac{\varepsilon_2\alpha_1^0}{\varepsilon_1\alpha_2^0}\right| < 1$ and $B_2/A_2 > 0$ [see Eq. (5.63)]. The dispersion relation of this mode is

$$\tanh\left(\frac{1}{2}\alpha_2^0 d\right) = -\frac{\varepsilon_2\alpha_1^0}{\varepsilon_1\alpha_2^0}. \tag{5.70}$$

The other equation corresponds to the odd mode, which satisfies the inequalities $\left|\frac{\varepsilon_2\alpha_1^0}{\varepsilon_1\alpha_2^0}\right| > 1$ and $B_2/A_2 < 0$ [see Eq. (5.63)]. And its dispersion relation is

$$\tanh\left(\frac{1}{2}\alpha_2^0 d\right) = -\frac{\varepsilon_1\alpha_2^0}{\varepsilon_2\alpha_1^0}. \tag{5.71}$$

From Eqs. (5.70) and (5.71), it is obvious as $d \to \infty$, both dispersion relations evolve into

$$\frac{\alpha_1^0}{\alpha_2^0} = -\frac{\varepsilon_1}{\varepsilon_2}. \tag{5.72}$$

Compared with Eq. (5.32), it is clear that both the even and odd modes reduce to the SPW propagating at the metal/dielectric interface. These two states and their propagation constant become degenerate for infinitely large d

$$\beta^0 = k_0\sqrt{\frac{\varepsilon_1\varepsilon_2}{\varepsilon_1+\varepsilon_2}}. \tag{5.73}$$

These two surface waves do not affect each other and travel freely when the thickness is very large, but as the film thickness decreases, they couple mutually and turn into even and odd modes. Let us consider the other limit of $d \to 0$, these two modes show different characteristics:

1. According to the dispersion relation Eq. (5.70) of even mode, there is $\alpha_1^0 = 0$, which corresponds to $\beta^0 = \kappa_0\sqrt{\varepsilon_1}$.

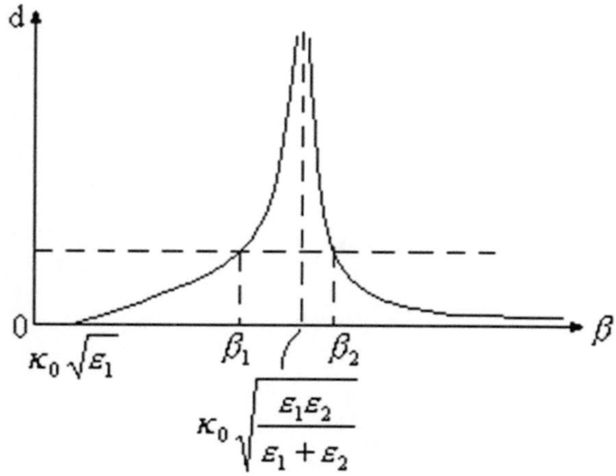

Fig. 5.11 Dispersion relation for the SPW on a symmetric metal film structure

2. Based on the dispersion relation Eq. (5.71) of the odd mode, both α_1^0 and α_2^0 approach infinite, so that $\beta^0 \to \infty$.

In view of the above analysis and the dispersion relation plotted in Fig. 5.11, clearly for a metal film of finite thickness d both even and odd modes will have their own propagation constants: β_1^0 for the even mode and β_2^0 for the odd mode. The allowed range for the even mode is

$$k_0\sqrt{\varepsilon_1} < \beta_1^0 < k_0\sqrt{\frac{\varepsilon_1\varepsilon_2}{\varepsilon_1+\varepsilon_2}}, \tag{5.74}$$

and the allowed range for the odd mode is

$$k_0\sqrt{\frac{\varepsilon_1\varepsilon_2}{\varepsilon_1+\varepsilon_2}} < \beta_2^0 < \infty. \tag{5.75}$$

5.4.2 Loss

This section focuses on the loss feature of metal film structure. Due to the optical property of metal in the optical frequency range, first-order perturbation method can be applied to deal with this issue. The main idea is to view the actual metal film with complex permittivity as a perturbed version of an ideal metal film, which is characterized by a real permittivity. If the perturbation of the ideal system's propagation constant is derived, its imaginary component represents the loss factor of the actual system.

1. Asymmetric metal film structure

According to the last section, when $\varepsilon_2 = \varepsilon_{r2} + i\varepsilon_{i2}$, the dispersion relation of the asymmetric metal film is

$$\tanh\left(\alpha_2' d\right) = -\frac{\dfrac{\varepsilon_2\alpha_1}{\varepsilon_1\alpha_2'} + \dfrac{\varepsilon_2\alpha_3}{\varepsilon_3\alpha_2'}}{1 + \dfrac{\varepsilon_2\alpha_1}{\varepsilon_1\alpha_2'} \cdot \dfrac{\varepsilon_2\alpha_3}{\varepsilon_3\alpha_2'}}, \tag{5.76}$$

where

$$\begin{aligned} \alpha_2' &= \sqrt{\beta^2 - k_0^2(\varepsilon_{r2} + i\varepsilon_{i2})} \\ &\approx \sqrt{\beta^2 - k_0^2\varepsilon_{r2}} - i\varepsilon_{i2}\frac{k_0^2}{2\sqrt{\beta^2 - k_0^2\varepsilon_{r2}}} \\ &= \alpha_2 - i\varepsilon_{i2}\frac{k_0^2}{2\alpha_2} \\ &= \alpha_2 - \Delta\alpha_2. \end{aligned} \tag{5.77}$$

For convenience, Eq. (5.76) is recasted into the following form

$$\alpha_2' d + \phi_{21}' + \phi_{23}' = 0, \tag{5.78}$$

where ϕ_{21}' and ϕ_{23}' are defined as

$$\tanh\phi_{21}' = \frac{\varepsilon_2\alpha_1}{\varepsilon_1\alpha_2'}, \tag{5.79}$$

$$\tanh\phi_{23}' = \frac{\varepsilon_2\alpha_3}{\varepsilon_3\alpha_2'}. \tag{5.80}$$

Using Eq. (5.77), it follows

$$\tanh\phi_{21}' \cong \frac{\varepsilon_{r2}\alpha_1}{\varepsilon_1\alpha_2}\left(1 + i\varepsilon_{i2}\frac{2\alpha_2^2 + k_0^2\varepsilon_{r2}}{2\alpha_2^2\varepsilon_{r2}}\right) = \tanh\phi_{21} + i\varepsilon_{i2}\frac{2\alpha_2^2 + k_0^2\varepsilon_{r2}}{2\alpha_2^2\varepsilon_{r2}} \cdot \frac{\varepsilon_{r2}\alpha_1}{\varepsilon_1\alpha_2}. \tag{5.81}$$

Similarly, it is easy to derive

$$\tanh\phi_{23}' \cong \frac{\varepsilon_{r2}\alpha_3}{\varepsilon_3\alpha_2}\left(1 + i\varepsilon_{i2}\frac{2\alpha_2^2 + k_0^2\varepsilon_{r2}}{2\alpha_2^2\varepsilon_{r2}}\right) = \tanh\phi_{23} + i\varepsilon_{i2}\frac{2\alpha_2^2 + k_0^2\varepsilon_{r2}}{2\alpha_2^2\varepsilon_{r2}} \cdot \frac{\varepsilon_{r2}\alpha_3}{\varepsilon_3\alpha_2}. \tag{5.82}$$

Consequently, we have

$$\begin{aligned}\Delta\phi_{21} &= \frac{\tanh\phi'_{21} - \tanh\phi_{21}}{1 - \tanh^2\phi_{21}} \\ &= i\varepsilon_{i2}\frac{2\alpha_2^2 + k_0^2\varepsilon_{r2}}{2\alpha_2} \cdot \frac{\varepsilon_1\alpha_1}{\varepsilon_1^2\alpha_2^2 - \varepsilon_{r2}^2\alpha_1^2},\end{aligned} \tag{5.83}$$

$$\begin{aligned}\Delta\phi_{23} &= \frac{\tanh\phi'_{23} - \tanh\phi_{23}}{1 - \tanh^2\phi_{23}} \\ &= i\varepsilon_{i2}\frac{2\alpha_2^2 + k_0^2\varepsilon_{r2}}{2\alpha_2} \cdot \frac{\varepsilon_3\alpha_3}{\varepsilon_3^2\alpha_2^2 - \varepsilon_{r2}^2\alpha_3^2}.\end{aligned} \tag{5.84}$$

Finally, the perturbed dispersion relation can be rewritten as

$$\alpha_2 d + \phi_{21} + \phi_{23} = \Delta\alpha_2 \cdot d - \Delta\phi_{21} - \Delta\phi_{23}, \tag{5.85}$$

and the dispersion relation for the ideal metal film structure Eq. (5.68) can be modified as

$$\alpha_2^0 d + \phi_{21}^0 + \phi_{23}^0 = 0. \tag{5.86}$$

Subtracting Eq. (5.85) from (5.86) and using differential formula, we get the perturbed propagation constant

$$\Delta\beta = i\varepsilon_{i2}\frac{k_0^2 d_{\text{loss}}}{2\beta d_{\text{eff}}}, \tag{5.87}$$

where the effective thickness of the metal film is defined as

$$d_{\text{eff}} = d + \frac{\varepsilon_{r2}\varepsilon_1\left(\alpha_2^2 - \alpha_1^2\right)}{\alpha_1\left(\varepsilon_1^2\alpha_2^2 - \varepsilon_{r2}^2\alpha_1^2\right)} + \frac{\varepsilon_{r2}\varepsilon_3\left(\alpha_2^2 - \alpha_3^2\right)}{\alpha_3\left(\varepsilon_3^2\alpha_2^2 - \varepsilon_{r2}^2\alpha_3^2\right)}, \tag{5.88}$$

and the loss length is given by the equation below

$$d_{\text{loss}} = d - \frac{2\alpha_2^2 + k_0^2\varepsilon_{r2}}{k_0^2}\left(\frac{\varepsilon_1\alpha_1}{\varepsilon_1^2\alpha_2^2 - \varepsilon_{r2}^2\alpha_1^2} + \frac{\varepsilon_3\alpha_3}{\varepsilon_3^2\alpha_2^2 - \varepsilon_{r2}^2\alpha_3^2}\right). \tag{5.89}$$

It is worth to point out that all the parameters in Eqs. (5.87) to (5.89) are from the ideal system, while the superscript (0) is omitted for clarity. This fact shows that the perturbed parameters of the actual metal film structure can be solved by the three equations above once the propagation constant of the ideal system is calculated via Eq. (5.68). From the expression (5.87) of the perturbed propagation constant, it is clear that $\Delta\beta$ is purely imaginary, which illustrates that the imaginary part of the complex metal permittivity can only effect the system's loss under the first-order perturbation approximation. This conclusion is consistent with that in Sect. 5.2.

2. Symmetric metal film structure

In the case of symmetric structure, i.e., $\varepsilon_1 = \varepsilon_3$, the dispersion relation of the perturbed system can be simplified as

$$\tanh\left(\alpha_2' d\right) = -\frac{2\frac{\varepsilon_2\alpha_1}{\varepsilon_1\alpha_2'}}{1+\left(\frac{\varepsilon_2\alpha_1}{\varepsilon_1\alpha_2'}\right)^2}. \tag{5.90}$$

The dispersion relation of the ideal system is given by Eq. (5.68) and the perturbed propagation constant is still presented by Eq. (5.87), but the definitions in Eqs. (5.88) and (5.89) should be replaced by

$$d_{\text{eff}} = d + \frac{2\varepsilon_{r2}\varepsilon_1\left(\alpha_2^2 - \alpha_1^2\right)}{\alpha_1\left(\varepsilon_1^2\alpha_2^2 - \varepsilon_{r2}^2\alpha_1^2\right)}, \tag{5.91}$$

$$d_{\text{loss}} = d - \frac{2\alpha_2^2 + k_0^2\varepsilon_{r2}}{k_0^2} \cdot \frac{2\varepsilon_1\alpha_1}{\varepsilon_1^2\alpha_2^2 - \varepsilon_{r2}^2\alpha_1^2}. \tag{5.92}$$

In order to study the loss characteristics of the odd and even modes, considering the limit case as $d \to 0$ and inserting Eqs. (5.91) and (5.92) into Eq. (5.87), it leads to

$$\begin{aligned}\Delta\beta &\cong -i\varepsilon_{i2}\frac{\alpha_1^2\left(2\alpha_2^2 + k_0^2\varepsilon_{r2}\right)}{2\beta\varepsilon_{r2}\left(\alpha_2^2 - \alpha_1^2\right)} \\ &= -i\varepsilon_{i2}\frac{\alpha_1^2\left(2\alpha_2^2 + k_0^2\varepsilon_{r2}\right)}{2\beta\varepsilon_{r2}\cdot k_0^2\left(\varepsilon_1 - \varepsilon_{r2}\right)}.\end{aligned} \tag{5.93}$$

Based on the analysis of Sect. 5.4.1, two important conclusions can be drawn:

1. For the even mode, $\alpha_1 \to 0$ when $d \to 0$; and from Eq. (5.93), we find $\Delta\beta \to 0$. So in the extremely thin metal film structure, this even mode with very small loss can travel a relatively long distance, and that is the reason why this mode is called as LRSPW. Although the LRSPW can propagate a longer distance than the ordinary SPW, its propagation distance is only of millimeter scale.
2. For the odd mode, both α_1 and α_2 approach infinite as $d \to 0$. Consequently, $\Delta\beta \to \infty$ indicates that the loss is infinitely large. So the odd mode cannot be excited in the limit of extremely thin metal film.

5.4.3 Excitation of the LRSPW

LRSPW can also be excited via a prism of high dielectric constant, Sarid [7] first proposed that it is theoretically possible to excite LRSPW via a symmetric structure ($\varepsilon_0 = \varepsilon_2$) as shown in Fig. 5.8. Soon Deck et al. [8] excited this kind of LRSPW via the ATR technology experimentally.

As shown in Fig. 5.12, the prism has a permittivity of ε_3, and a metal film of permittivity $\varepsilon_1 = \varepsilon_{r1} + i\varepsilon_{i1}$ and thickness d_1 is sandwiched between two dielectrics of ε_0 and ε_2. The thickness of dielectric 2 is d_2, and dielectric 0 is a half-infinite medium. The incident angle of a TM polarized light at the bottom interface of prism is θ_3. So the magnetic field distribution of the LRSPW is

$$H_y(x) = \begin{cases} A_3 e^{\alpha_3(x-d_2)} + B_3 e^{-\alpha_3(x-d_2)}, & (x > d_2), \\ A_2 e^{\alpha_2 x} + B_2 e^{-\alpha_2 x}, & (0 < x < d_2), \\ A_1 e^{\alpha_1 x} + B_1 e^{-\alpha_1 x}, & (-d_1 < x < 0), \\ A_0 e^{\alpha_0(x+d_1)}, & (x < -d_1), \end{cases} \tag{5.94}$$

where

$$\begin{aligned} &\alpha_j = \left(\beta^2 - k_0^2 \varepsilon_j\right)^{1/2}, \quad (j = 0, 1, 2, 3), \\ &\beta = k_0 \sqrt{\varepsilon_3} \sin\theta_3. \end{aligned} \tag{5.95}$$

The oscillating field in the prism results in an imaginary parameter α_3 in that region. α_1 is complex due to the complex permittivity ε_1 of metal, and both parameters of α_0 and α_2 are real. Using the boundary condition, the expression of reflectivity is

$$R = \left|\frac{B_3}{A_3}\right|^2 = \left|\frac{\gamma_{23} - \gamma_{012} e^{-2\alpha_2 d_2}}{1 + \gamma_{23}\gamma_{012} e^{-2\alpha_2 d_2}}\right|^2, \tag{5.96}$$

where

$$\gamma_{012} = \frac{\gamma_{12} + \gamma_{01} e^{-2\alpha_1 d_1}}{1 + \gamma_{12}\gamma_{01} e^{-2\alpha_1 d_1}}, \tag{5.97}$$

$$\gamma_{23} = \frac{\varepsilon_2\alpha_3 - \varepsilon_3\alpha_2}{\varepsilon_2\alpha_3 + \varepsilon_3\alpha_2}, \tag{5.98}$$

and

$$\gamma_{01} = \frac{\varepsilon_0\alpha_1 - \varepsilon_1\alpha_0}{\varepsilon_0\alpha_1 + \varepsilon_1\alpha_0}, \tag{5.99}$$

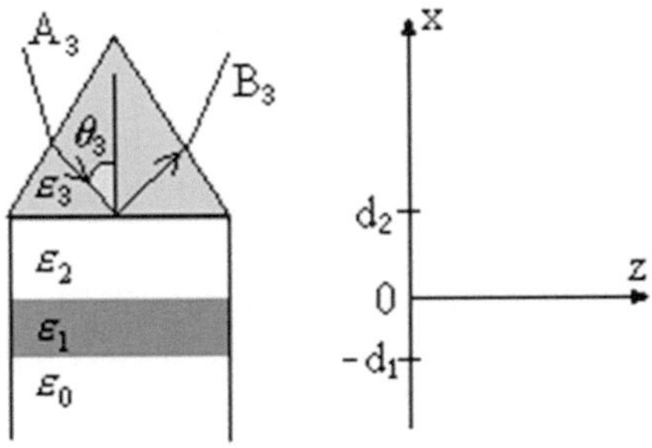

Fig. 5.12 Experimental setup for LRSPW excitation

$$\gamma_{12} = \frac{\varepsilon_1\alpha_2 - \varepsilon_2\alpha_1}{\varepsilon_1\alpha_2 + \varepsilon_2\alpha_1}. \tag{5.100}$$

For the experimental setup in Fig. 5.12, once the incident wavelength and structure parameters are fixed, the reflectivity curve as a function of incident angle θ_3 can be plotted via Eq. (5.96). A concrete example is presented in Fig. 5.13, the related parameters are $\lambda = 632.8\,\text{nm}$, $\varepsilon_0 = \varepsilon_2 = 2.295$, $\varepsilon_3 = 3.24$, $\varepsilon_1 = -18.0 + i0.7$, $d_1 = 17\,\text{nm}$, and $d_2 = 1200\,\text{nm}$. The resonance dip is very sharp, indicating that the loss of LRSPW is very small and this wave can travel a longer distance. Compared with the ATR spectrum of the SPW in Fig. 5.6, it is obvious that the width of the resonance dip of SPW is almost one order of magnitude of that of the LRSPW.

5.4.4 Field Enhancement Effect of the LRSPW

Different from the SPW propagates along a single interface of metal, the LRSPW is generated by the coupling between the two SPW excited on both interfaces of a metal film. This coupling decreases the proportion of the energy that dwells inside

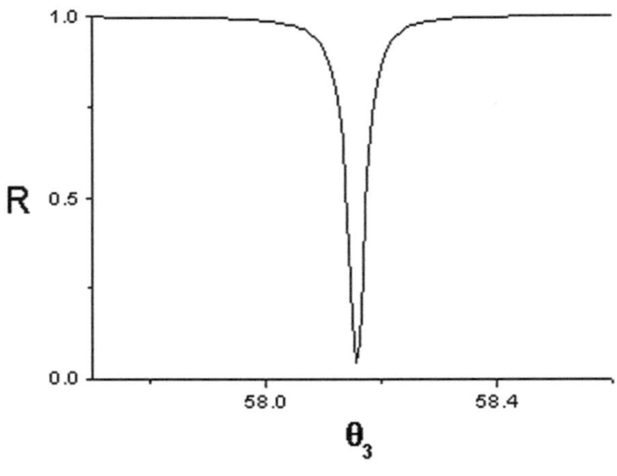

Fig. 5.13 The ATR spectrum of the LRSPW

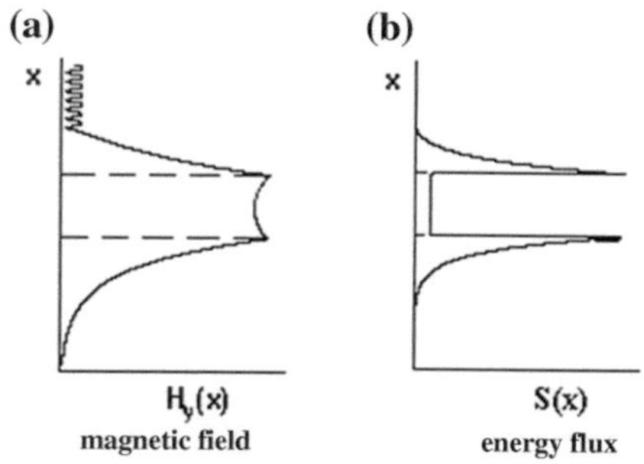

Fig. 5.14 The magnetic field (**a**) and Poynting vector (energy flux), (**b**) distribution of the LRSPW

the metal and drastically enhances the field intensity at the metal boundary. Figure 5.14a, b plots the distributions of the magnetic field and Poynting vector (energy flux) of the LRSPW, respectively. From these figures, we can see that the energy almost decreases to zero inside the metal, which reduces the loss and enable a long-distance transport.

5.5 Determination of Thickness and Permittivity of Thin Metal Films via a Modified ATR Configuration

Besides the double-wavelength method in Sect. 5.3, which was applied to measure the thickness and permittivity of a thin metal film, we proposed another method via exchanging the medium adjacent to the metal film [5]. Different from using two wavelengths to excite the SPW, our method uses only one wavelength to excite two different kinds of SPW in the ATR configurations, where the metal film is covered by different dielectrics. Both methods work very well and require two measurements, which may introduce more uncertainty in the experiment, e.g., the variation of the temperature, laser intensity, and the error of the angular scan. So in this section, we propose a new ATR configuration, in which both SPW and modified LRSPW can be excited. The resulted reflection spectrum has two resonance dips corresponding to these two kinds of surface waves, which enable the single angular scan measurement become possible [9]. This new method can avoid the dispersion problem of the double-wavelength method and eliminate the uncertainty between two measurements.

5.5.1 Measurement Principle

As shown in Fig. 5.15, the modified ATR structure consists of the prism with a high refractive index, an polymer film and the metal film, which is laid on the base side

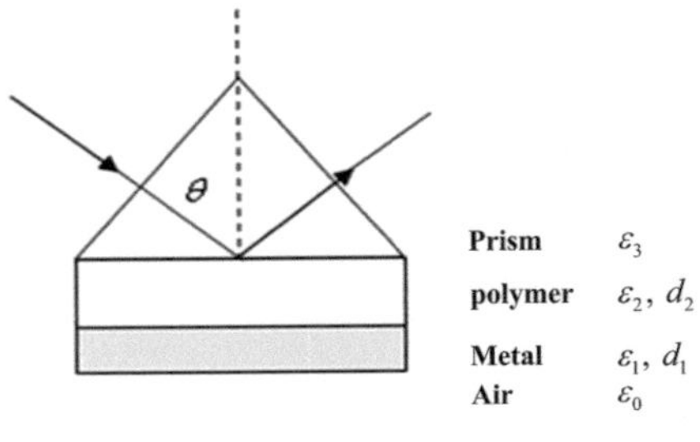

Fig. 5.15 Modified ATR configuration

of polymer film. The extra polymer film has a refractive index $\sqrt{\varepsilon_2}$, which is smaller than the prism's $\sqrt{\varepsilon_3}$. When the laser passes through the prism and enters into the polymer film, and its wave vector component along the interface matches the SPW's wave vector at the metal/air interface, the SPW can be excited. This is similar with the case of the traditional Kretschmann's structure. When the incident angle $\theta > \sin^{-1}(\varepsilon_2/\varepsilon_3)^{1/2}$, total reflection occurs at the prism/polymer interface and an evanescent field generates in the polymer film. Meanwhile, modified LRSPW can be excited on both sides of the metal film. However, the LRSPWs are excited in the symmetric structure where the metal film is sandwiched between the same medium, and the configuration in Fig. 5.15 is not symmetric, that is the reason why this kind of surface wave is called as the modified LRSPW.

The reflectivity expression of the modified ATR structure can be derived by the Fresnel formula

$$R = \left| \frac{r_{32} + r_{210} \exp(-2\alpha_2 d_2)}{1 + r_{32} r_{210} \exp(-2\alpha_2 d_2)} \right|^2, \tag{5.101}$$

where

$$r_{210} = \left| \frac{r_{21} + r_{10} \exp(-2\alpha_1 d_1)}{1 + r_{21} r_{10} \exp(-2\alpha_1 d_1)} \right|^2, \tag{5.102}$$

$$\begin{cases} r_{21} = \dfrac{\varepsilon_1 \alpha_2 - \varepsilon_2 \alpha_1}{\varepsilon_1 \alpha_2 + \varepsilon_2 \alpha_1}, \\ r_{10} = \dfrac{\varepsilon_0 \alpha_1 - \varepsilon_1 \alpha_0}{\varepsilon_0 \alpha_1 + \varepsilon_1 \alpha_0}, \\ r_{32} = \dfrac{\varepsilon_2 \kappa_3 - i\varepsilon_3 \alpha_2}{\varepsilon_2 \kappa_3 + i\varepsilon_3 \alpha_2}, \end{cases} \tag{5.103}$$

$$\begin{cases} \alpha_j = \left(\beta^2 - k_0^2 \varepsilon_j\right)^{1/2}, \quad (j = 0, 1, 2), \\ \kappa_3 = \left(k_0^2 \varepsilon_3 - \beta^2\right)^{1/2}, \end{cases} \tag{5.104}$$

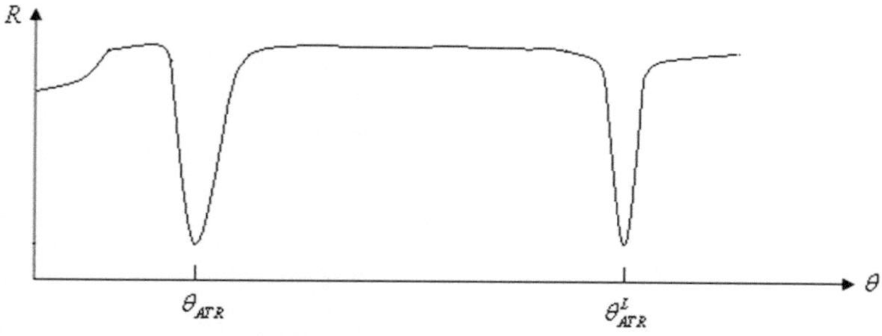

Fig. 5.16 The reflection spectrum of the modified ATR configuration

and

$$\beta = k_0\sqrt{\varepsilon_3}\sin\theta. \tag{5.105}$$

When the incident angle θ matches one of the resonance angles of the two surface waves, the corresponding surface waves can be excited. We use θ_{ATR} and θ^L_{ATR} to represent the resonance angle of the SPW and the modified LRSPW in this ATR configuration. The reflectivity spectrum calculated by Eq. (5.101) is plotted in Fig. 5.16, where the first resonance dip is due to the traditional SPW, and the second dip corresponds to the modified LRSPW. From the resonance dip of the SPW, information of the resonance angle θ_{ATR}, the half width of the dip W_θ, and the reflectivity minimum $R_{\min}$ can be measured. By the numerical fitting, two sets of solutions on the metal film thickness and permittivity can be obtained. Additional information must be provided to choose the correct set of solution. While double-wavelength method uses another wavelength and the exchanging media method adopts another dielectric, this method produces an extra resonance dip of the modified LRSPW. And the correct set of solution can be easily picked up by numerically fitting the θ^L_{ATR}.

5.5.2 Experiment and Measurement

Let us see a concrete example, the parameters of the sample are

$$\varepsilon_0 = 1.0, \quad \varepsilon_2 = 2.2201, \quad \varepsilon_3 = 3.0625,$$

$$d_2 = 0.72\,\mu\mathrm{m}, \quad \lambda = 0.788\,\mu\mathrm{m}.$$

The laser of $0.788\,\mu\mathrm{m}$ is applied to the angular scan of the sample, which generates the ATR spectrum as shown in Fig. 5.16. Using the conventional resonance dip of the SPW, two sets of solutions can be calculated.

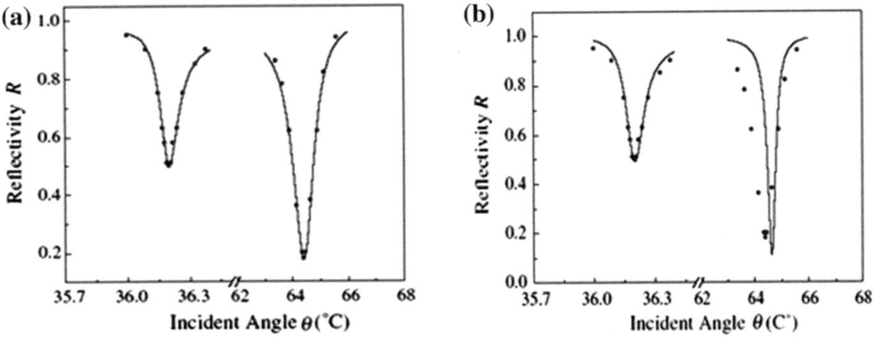

Fig. 5.17 Comparison between the numerical simulation and experimental results. **a** the first set solution, **b** the second set solution

$$1.\ \varepsilon_1 = -26.5 + i1.47, \quad d = 68.5\text{nm},$$
$$2.\ \varepsilon_1 = -27.6 + i0.24, \quad d = 46.3\text{nm}$$

Let's insert these two sets of solutions into the reflectivity expression Eq. (5.101), and compare the two sets of numerical simulations with the experimental measured data, which are plotted in Fig. 5.17. The comparison of the first set of $\varepsilon_1 = -26.5 + i1.47, d = 68.5$ nm is shown in Fig. 5.17a, where the solid line denotes the numerical simulated reflectivity. Figure 5.17b corresponds to the second set of solution.

It is clear from the above figure that the first set solution is true. It should be pointed out here in order to effectively excite the modified LRSPW, the thickness of the polymer film must be chosen carefully. In this example, numerical simulation shows that the LRSPW can be excited when the thickness of the polymer film is in the range of $0.5 \sim 0.8$ μm.

References

1. X. She, *Physical basis of guide wave optics* (Beijing Jiaotong University Press, Beijing, 2002). (in Chinese)
2. A. Otto, Excitation of nonradiative surface plasma waves in sliver by the method of frustrated total reflection. Z. Physik **216**, 38 (1968)
3. E. Kretschmann, H. Raether, Radiative decay of non radiative surface plasmon excited by light. Z. F. Naturforschung A **23**, 2135 (1968)
4. W.P. Chen, J.M. Chen, Use of surface plasma waves for determination of the thickness and optical constants of thin metallic films. J. Opt. Soc. Am. **71**, 189 (1981)
5. F. Yang, Z. Cao, J. Fang, Use of exchanging media in ATR configurations for determination of thickness and optical constants of thin metallic films. Appl. Opt. **27**, 11 (1988)
6. T. Lopez-Rios, G. Vuye, In situ investigation of metallic surfaces by surface plasmon ATR spectroscopy, electrical resistance measurements and Auger spectroscopy. J. Phys. **15**, 456 (1982)
7. D. Sarid, Long-range surface-plasma waves on very thin metal film. Phys. Rev. Lett. **47**, 1981 (1927)

8. R.T. Deck, D. Sarid, Enhancement of second-harmonic generation by coupling to long-range surface plasmons. J. Opt. Soc. Am. **72**, 1613 (1982)
9. Y. Ding, Z. Cao, Q. Shen, Determination of optical waveguide refractive-index profiles with the inverse analytic transfer matrix method. Opt. Quant. Electron. **35**, 1091 (2003)

Chapter 6
Symmetrical Metal-Cladding Waveguide

Abstract In this chapter, we first briefly discuss the dispersion equation of the symmetrical metal-cladding waveguide (SMCW), and review some particular properties of the TM_0 mode and TM_1 mode excited in the SMCW. Then, we give the principle of the free-space coupling technology and the characteristics of the ultrahigh-order mode. Finally, two amusing optical phenomena observed in the SMCW, i.e., the wideband slow light effect and the conical reflection, are analyzed in detail.

Keywords Symmetrical metal-cladding waveguide · Dispersion equation · Free-space coupling technology · Ultrahigh-order mode · Wideband slow light · Conical reflection

In Chap. 5, we have discussed the surface plasmon wave excited along the interface between a metal and a non-absorb dielectric, the long-range surface plasmon wave propagated in a thin metal film, and several practical applications of these two surface modes. However, since the metal is a strongly absorbing material, the propagation lengths of the above modes are not large enough to construct the integrated optical devices for the purpose of optical communication. Nevertheless, the metal is frequently employed as a cladding layer in some waveguide structures. For example, the metal–dielectric–metal structure can function as a negative refraction lens [1], which is important for the manipulation of optical waves in the nanoscale. Moreover, the metal film can be used as an electrode in the electro-optic modulator and as a mirror to reflect or enlarge the electromagnetic field along the metal–dielectric interface. Therefore, the existence of metal would largely change the dispersion relation and bring many unique optical properties, which are different from those of the dielectric planar waveguides.

X. Wang et al., *Progress in Planar Optical Waveguides*,
Springer Tracts in Modern Physics 266, DOI 10.1007/978-3-662-48984-0_6

6.1 Dispersion Equation

6.1.1 Dispersion Properties

The configuration of the SMCW is illustrated in Fig. 6.1. A dielectric slab with thickness d and dielectric constant ε_1 acting as the guiding layer is sandwiched by two metal films, which serve as the coupling layer and the substrate, respectively. By denoting the dielectric constant of metal film as ε_2 and ignoring its imaginary part, the dispersion equation of the SMCW for TE mode is written as follows:

$$\kappa_1 \mathrm{d} = m\pi + 2\arctan\left(\frac{\alpha_2}{\kappa_1}\right). \tag{6.1}$$

and for TM mode:

$$\kappa_1 \mathrm{d} = m\pi + 2\arctan\left(\frac{\varepsilon_1\alpha_2}{\varepsilon_2\kappa_1}\right). \tag{6.2}$$

where

$$\begin{cases} \kappa_1 = \left(k_0^2\varepsilon_1 - \beta^2\right)^{1/2} \\ \alpha_2 = \left(\beta^2 - k_0^2\varepsilon_2\right)^{1/2} \end{cases}. \tag{6.3}$$

Several important conclusions can be summarized from the above dispersion equations.

(i) The effective refractive index (RI) of the guided mode is bounded by

$$0 < \beta/k_0 < \sqrt{\varepsilon_1}, \tag{6.4}$$

which is larger than the corresponding range of the dielectric planar waveguide and the metal clad leaky waveguide [2].

(ii) The TM_0 and TM_1 modes have some particular properties. More details will be given in the next subsection.

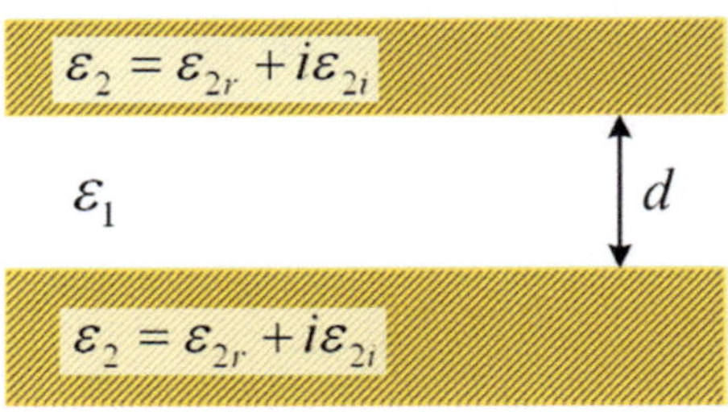

Fig. 6.1 Configuration of the SMCW

(iii) For the given light frequency ω and the thickness of guiding layer d, the propagation constant β of the *m*th TE mode is smaller than that of the *m*th TM mode. This property of the SMCW is identical with that of the metal clad leaky waveguide, but is different from that of the dielectric planar waveguide.
(iv) Besides the TM_0 mode, other excited modes have each own cutoff thickness d_s. When the mode is cut off, the propagation constant tends to zero, i.e., $\beta \to 0$, for TE mode:

$$k_0 d_s^{TE_m} = \frac{m\pi}{\sqrt{\varepsilon_1}} + \frac{2}{\sqrt{\varepsilon_1}} \arctan\sqrt{-\frac{\varepsilon_{r2}}{\varepsilon_1}}, \tag{6.5}$$

and for TM mode:

$$k_0 d_s^{TM_m} = \frac{m\pi}{\sqrt{\varepsilon_1}} - \frac{2}{\sqrt{\varepsilon_1}} \arctan\sqrt{-\frac{\varepsilon_1}{\varepsilon_{r2}}}. \tag{6.6}$$

Clearly, for the same mode order, we obtain

$$d_s^{TE_m} > d_s^{TM_m}. \tag{6.7}$$

But for the dielectric planar waveguide, the cutoff thickness of the same mode order is determined by

$$k_0 d_s^{TE_m} = k_0 d_s^{TM_m} = \frac{m\pi}{\sqrt{\varepsilon_1 - \varepsilon_2}}. \tag{6.8}$$

By setting $m = 0$, Eq. (6.8) is recast into

$$d_s^{TE_0} = d_s^{TM_0} = 0. \tag{6.9}$$

It implies that there is no cutoff thickness for the TM_0 and TM_1 modes of the dielectric planar waveguide.

Moreover, by using the identical equation

$$\arctan(|u|) = \frac{\pi}{2} - \arctan(|u|^{-1}), \tag{6.10}$$

and comparing Eq. (6.5) with Eq. (6.6), it is easy to get

$$d_s^{TE_m} = d_s^{TM_{m+1}}, \tag{6.11}$$

which indicates that the cutoff thicknesses of the *m*th TE mode and the $(m + 1)$th TM mode in the SMCW structure are identical.

6.1.2 TM_0 Mode and TM_1 Mode

1. TM_0 mode

By substituting $m = 0$ into Eq. (6.2), we have

$$\kappa_1 \mathrm{d} = 2\arctan\left(\frac{\varepsilon_1\alpha_2}{\varepsilon_2\kappa_1}\right). \tag{6.12}$$

Several key conclusions can be summarized here:

(i) Since the real part of ε_2 is negative, Eq. (6.12) has no solution in the range of $\beta/\mathrm{k}_0 < \sqrt{\varepsilon_1}$.

(ii) In the case of $\beta/\mathrm{k}_0 > \sqrt{\varepsilon_1}$, κ_1 can be represented as $i\alpha_1$. As a result, the oscillating wave in the guiding layer is replaced by a superposition of two exponential decay evanescent waves and Eq. (6.12) is recast into

$$\tanh\left(\frac{1}{2}\alpha_1 \mathrm{d}\right) = -\frac{\varepsilon_1\alpha_2}{\varepsilon_2\alpha_1}. \tag{6.13}$$

When $\mathrm{d} \to \infty$, we get

$$-\frac{\varepsilon_1\alpha_2}{\varepsilon_2\alpha_1} = 1, \tag{6.14}$$

which is a surface plasmon wave excited along the interface between the upper metal layer and the dielectric slab, and the effective RI is expressed as follows:

$$\frac{\beta}{k_0} = \sqrt{\frac{\varepsilon_1\varepsilon_2}{\varepsilon_1 + \varepsilon_2}} > \sqrt{\varepsilon_1}. \tag{6.15}$$

As $\mathrm{d} \to 0$, it is easy to prove $\alpha_1 \to \infty$ and $\alpha_2 \to \infty$, and then

$$\beta/k_0 \to \infty.$$

Thus, it is clear that the effective RI of the TM_0 mode is bounded by

$$\sqrt{\frac{\varepsilon_1\varepsilon_2}{\varepsilon_1 + \varepsilon_2}} < \frac{\beta}{k_0} < \infty, \tag{6.16}$$

which is out of the range of guided modes (see Eq. (6.4)). Therefore, we can conclude that the TM_0 mode is a surface mode.

2. TM_1 mode

By using the identical Eq. (6.10), the dispersion equation of TM_1 mode can be rewritten as follows:

$$\kappa_1 \mathrm{d} = 2\arctan\left(-\frac{\varepsilon_2 \kappa_1}{\varepsilon_1 \alpha_2}\right). \tag{6.17}$$

Clearly, it can be solved in the range of $0 < \beta/k_0 < \sqrt{\varepsilon_1}$. In this case, the TM_1 mode is a guided mode.

By setting $\beta/k_0 = 0$, the cutoff thickness of the TM_1 mode is as follows:

$$\mathrm{d}_s^{TM_1} = \frac{2}{k_0\sqrt{\varepsilon_1}} \arctan\sqrt{-\frac{\varepsilon_2}{\varepsilon_1}}. \tag{6.18}$$

If one sets $\beta/k_0 = \sqrt{\varepsilon_1}$, the critical thickness of the TM_1 mode can be derived from Eq. (6.17)

$$\mathrm{d}_c^{TM_1} = -\frac{2\varepsilon_2}{k_0 \varepsilon_1 \sqrt{\varepsilon_1 - \varepsilon_2}}. \tag{6.19}$$

For the case of $\beta/k_0 > \sqrt{\varepsilon_1}$, there is $\kappa_1 = \mathrm{i}\alpha_1$, and Eq. (6.17) is recast into

$$\tanh\left(\frac{1}{2}\alpha_1 \mathrm{d}\right) = -\frac{\varepsilon_2 \alpha_1}{\varepsilon_1 \alpha_2}. \tag{6.20}$$

Combined with the discussions given in Chap. 5, we can conclude that the TM_0 mode expressed by Eq. (6.13) and the TM_1 mode expressed by Eq. (6.20) are the symmetric and antisymmetric superpositions of two surface plasmon waves excited in the thin metal film sandwiched by two identical dielectric layers, respectively.

When $\mathrm{d} \to \infty$, Eq. (6.20) is rewritten as follows:

$$-\frac{\varepsilon_2 \alpha_1}{\varepsilon_1 \alpha_2} = 1. \tag{6.21}$$

In this case, the effective RI is as follows:

$$\beta/k_0 = \sqrt{\frac{\varepsilon_1 \varepsilon_2}{\varepsilon_1 + \varepsilon_2}}. \tag{6.22}$$

According to Eqs. (6.15) and (6.22), it is found that both the TM_0 mode and the TM_1 mode will degenerate with each other on condition of $\mathrm{d} \to \infty$. From the physical perspective, if the thickness of the guiding layer is infinitely large, the surface plasmon waves excited at two metal–dielectric interfaces cannot coupled with each other because these two modes are the free-surface plasmon waves in the case of $d \rightarrow \infty$.

From the above discussion, it is easy to obtain that the effective RI of the TM_1 mode is bounded by

$$0 < \beta/k_0 < \sqrt{\frac{\varepsilon_1 \varepsilon_2}{\varepsilon_1 + \varepsilon_2}}. \tag{6.23}$$

Combined with Eqs. (6.16) and (6.23), the effective RI of the SMCW (including the guided modes and the surface mode) is bounded by

$$0 < \beta/k_0 < \infty. \tag{6.24}$$

The dispersion curves of the SMCW and the symmetrical dielectric waveguide are shown in Fig. 6.2a, b, respectively. The solid curves are for TM modes, and the dashed curves are for TE modes. In addition, the dispersion properties of the TM_0 mode and the TM_1 mode excited in the SMCW are listed in Table 6.1.

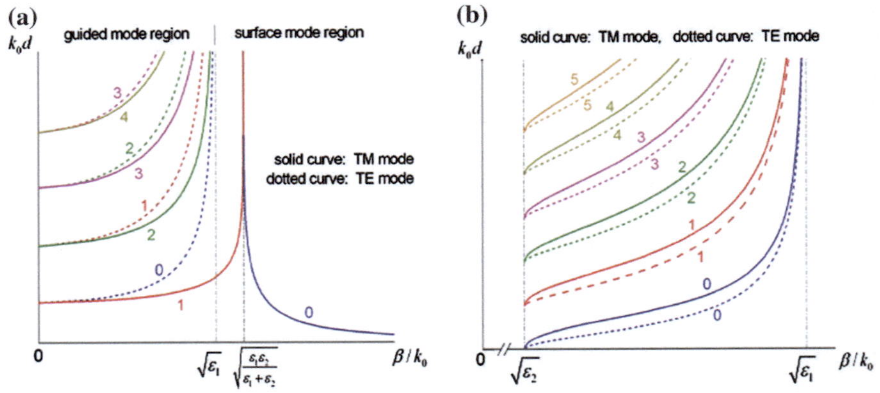

Fig. 6.2 Dispersion curves of (**a**) symmetrical metal-cladding waveguide and (**b**) symmetry dielectric waveguide

Table 6.1 The dispersion properties of the TM_0 mode and the TM_1 mode excited in the SMCW

β/k_0	d	Mode types
$0 < \beta/k_0 < \sqrt{\varepsilon_1}$	$d_s < d < d_c$	TM_1 is a guided mode
$\sqrt{\varepsilon_1} < \beta/k_0 < \sqrt{\frac{\varepsilon_1 \varepsilon_2}{\varepsilon_1 + \varepsilon_2}}$	$d_s < d < \infty$	TM_1 is a surface mode
$\sqrt{\frac{\varepsilon_1 \varepsilon_2}{\varepsilon_1 + \varepsilon_2}} < \beta/k_0 < \infty$	$0 < d < \infty$	TM_0 is a surface mode

6.1.3 *The Degeneracy of TM_0 Mode and TM_1 Mode*

As shown in Fig. 6.2a, the TM_0 mode and the TM_1 mode of the SMCW will be degenerated with each other when the thickness of the guiding layer is infinitely large. Actually, the degeneracy of these two modes will occur once the thickness of the guiding layer is increased beyond one certain finite value (defined as the degeneracy thickness). The reason is that the attenuated total reflection (ATR) dip has a nonzero resonance width, which is attributed to the existence of the intrinsic and radiative damping in the SMCW structure. To illustrate this issue, a demonstration experiment is given. A prism made from k9 glass (n = 1.507 @ $\lambda = 780$ nm) is coated with an about 50-nm-thick silver film ($\varepsilon_2 = -28 + i1.0$ @ $\lambda = 780$ nm) to serve as the coupling layer. A 200-nm-thick silver film is sputtered on a glass slab to act as the substrate. These two components are then clamped together to form an SMCW structure in which the middle air gap is employed as the guiding layer. The thickness of the air gap can be adjusted via changing the pressure of the clamper. The transformation from the separation state of the TM_0 mode and the TM_1 mode to their degeneracy state is shown in Figs. 6.3, 6.4, and 6.5.

In Fig. 6.5, the thickness of the guiding layer is already beyond the degeneracy thickness; it therefore can be confirmed that the first ATR dip from the right side is the degenerated dip of the TM_0 mode and the TM_1 mode; and the second and third resonant dips represent the TM_2 mode and the TM_3 mode, respectively. In Fig. 6.3, the TM_0 mode and the TM_1 mode are apart from each other since the thickness of the guiding layer is smaller than the degeneracy thickness; therefore, the three ATR dips from the right side to the left side represent the TM_0, TM_1, and TM_2 modes, respectively. Clearly, Fig. 6.4 illustrates the critical state of the degeneracy between the TM_0 mode and the TM_1 mode. Until here, we find that the mode order of the ATR spectrum can be exactly identified only after a careful distinguishing on the separation state and the degeneracy state of these two fundamental modes. More details about the identification of mode order have been given in [3].

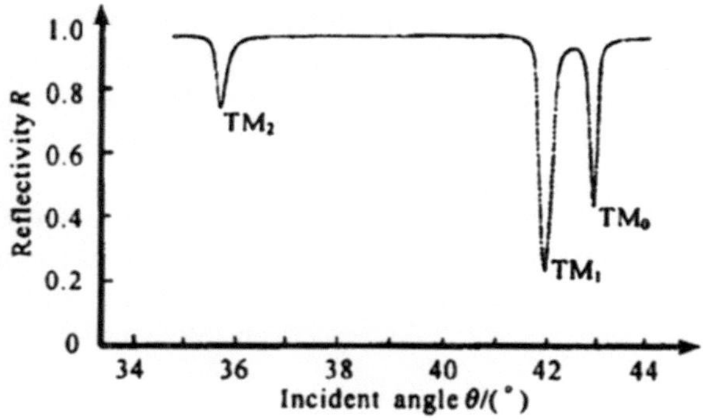

Fig. 6.3 The ATR curve with d = 1480 nm

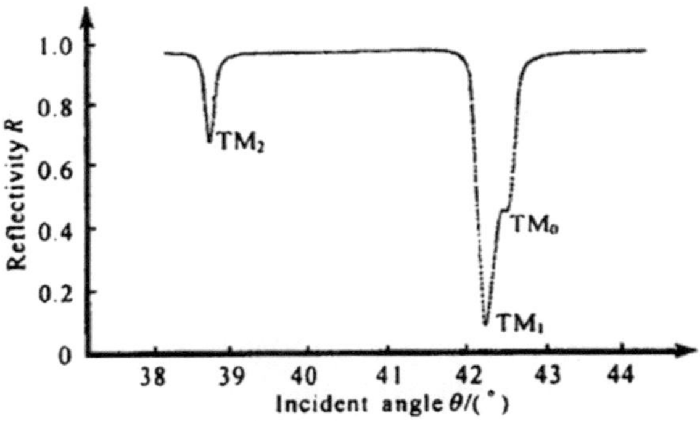

Fig. 6.4 The ATR curve with d = 2200 nm

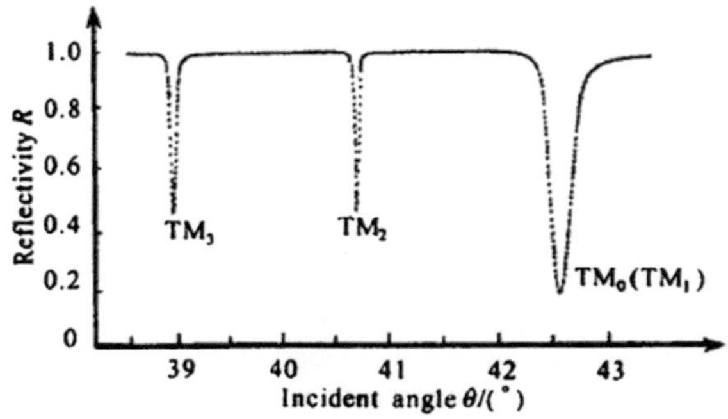

Fig. 6.5 The ATR curve with d = 3200 nm

6.2 Free-Space Coupling Technology

In the guided-wave optics, how to efficiently couple the light into the optical waveguide is an important issue and has attracted considerable interest. The commonly used coupling techniques are the end-face coupling [4], the prism coupling [5], the grating coupling [6], and the tapered film coupling [7]. In this subsection, we report on a new coupling technology [8], which can directly transfer energy of a light beam from the free space into a SMCW structure without employing any coupling components. In order to understand the physical mechanism of this new coupling technology, a full discussion of prism–waveguide coupling system is necessary.

The prism–waveguide coupling system is shown in Fig. 6.6. The prism–waveguide coupling system is in essence a system of four stratified layers, which are the substrate, the guiding layer, the cover, and the prism and denoted by the subscripts $j = 0, 1, 2, 3$, respectively. Without loss of any generality, we assume

$$n_1 > n_0 > n_2. \tag{6.25}$$

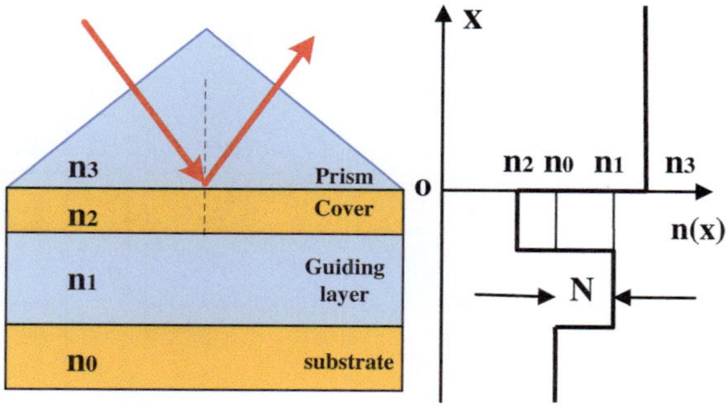

Fig. 6.6 The prism–waveguide coupling system

The effective RI is limited to the range

$$n_1 > N > n_0, \tag{6.26}$$

in which the light guiding is possible. If there is no prism and the light beam is directly coupled from the cover layer (the incident angle is assumed as θ_2), then the effective RI has the form

$$N = n_2 \sin\theta_2. \tag{6.27}$$

From Eqs. (6.25), (6.26), and (6.27), we get sin $\theta_2 > 1$. Therefore, it is impossible for the waveguide structure sketched in Fig. 6.6 to couple the light energy directly from the free space (here the cover layer is assumed to be air) into the guiding layer.

In order to couple the light beam into the guiding layer, the most well-known method is placing over the cover layer with a higher refractive index prism, that is,

$$n_3 > N. \tag{6.28}$$

Here, the light coupling is possible since the incident angle θ_3 satisfies sin $\theta_3 < 1$. In this case, the light fields propagate as oscillating waves in the prism and the guiding layer but as evanescent waves in the cover layer and the substrate. As described in Chap. 4, this prism–waveguide coupling process is called as an optical tunneling effect, whose quantum analogue is the phenomenon of the microparticle tunneling through a potential barrier.

As shown in Fig. 6.7, the thickness of the upper metal film is a finite value (denoted as S). We have known that the effective RI of the guided modes in the SMCW structure satisfies Eq. (6.4). When the light beam is directly incident from the free space ($n_3 = 1$), the modes with the effective RI that obeys $N < n_3$ can be excited. It means that the light beam can be coupled into the guiding layer of the SMCW directly from the free space without using any couple components. The physical mechanism of this new coupling technology can also be understood as following: When the beam is incident from the air to the upper metal film, it will generate an evanescent wave in the metal; since the upper metal film is thin

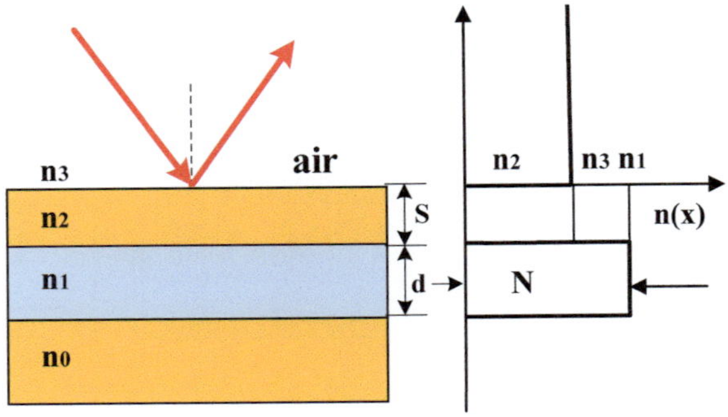

Fig. 6.7 The free-space coupling technology in the SMCW structure

(typically dozens of nanometer), the tail of evanescent wave can reach the metal-guiding layer interface, and then, it is reflected at this interface resulting in another evanescent wave with an opposite propagation direction; due to the interaction between these two evanescent waves, the incident light energy can be coupled into the waveguide.

The experimental demonstration of the free-space coupling technology is presented. A BKT glass flat with a thickness 106 μm is sandwiched between two gold films to form the SMCW structure, where the upper gold film with a thickness 46 nm acts as the coupling layer and the bottom gold film with a thickness 200 nm serves as the substrate, respectively. The dielectric constant of the glass and the gold is $\varepsilon_1 = 2.25$ and $\varepsilon_2 = -15 + i1.5$, respectively. A collimated light beam from a

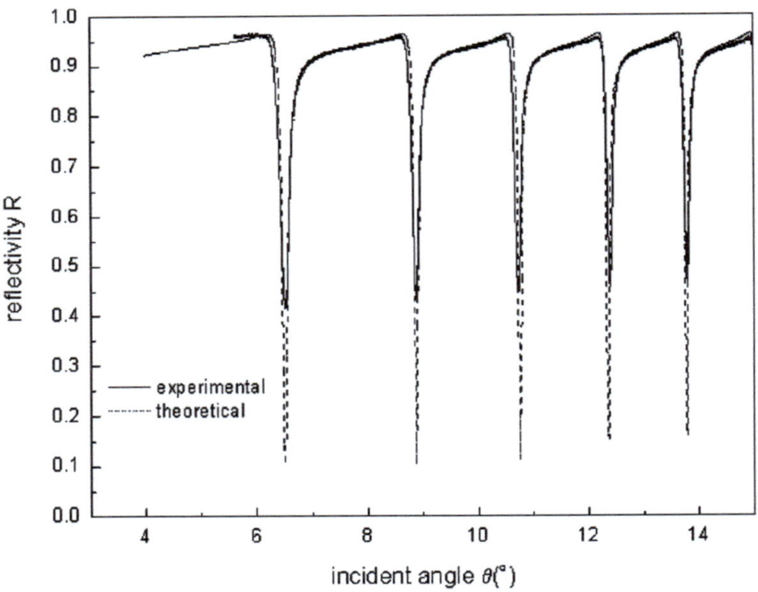

Fig. 6.8 ATR spectrum obtained from the SMCW

diode laser of 780 nm is incident on the top surface of the gold film, and the angular scan is carried out by a computer-controlled $\theta/2\theta$ goniometer. Experimental result of the free-space coupling is shown in Fig. 6.8 (solid curve). A series of dips in reflectivity spectrum due to the resonant transfer of energy into guided modes (TE) can be clearly observed. The theoretical (dashed curve) result is also given. It indicates that more than 50 % light energy has been fed into the optical waveguide.

6.3 Ultrahigh-Order Mode

When the thickness of the guiding layer reaches a millimeter scale, the SMCW can accommodate thousands of guided modes. For example, using the parameters: $\varepsilon_2 = -28 + i1.8$, $\varepsilon_1 = 2.278$, $S = 30$ nm, d = 0.38 mm, and $\lambda = 859.8$ nm, m is 1333 for the highest mode. When the light beam is coupled from the free space with a large incident angle, it is difficult to differentiate the adjacent guide modes because the mode density in the SMCW is extremely large and the corresponding ATR dips overlap with each other. On the other hand, a series of discrete guided modes can be excited at certain extremely small incident angles. These modes are called as the ultrahigh-order modes [9] in our terminology since the order of these modes is extremely large and their effective RI is small. From the ray-optical perspective, within the same longitudinal propagation distance, the ultrahigh-order mode is rebounded at the interfaces with more times and the ray propagation length is longer since the incident angles of the ultrahigh-order mode at the top and bottom interfaces of the guiding layer are extremely small. As a result, the ultrahigh-order modes hold many peculiar optical properties different from the low-order modes.

(i) The ATR dips of the ultrahigh-order modes are discrete.
On the basis of dispersion Eqs. (6.1) and (6.2) of the SMCW, we can obtain an approximate formula

$$\Delta m \propto \sin 2\theta_3 \cdot \Delta\theta_3. \tag{6.29}$$

Since $\Delta m = 1$, when the incident angle θ_3 is smaller, a bigger $\Delta\theta_3$ is obtained. So the ATR dips of the ultrahigh-order modes are discrete. Such property is convenient to design a comb filter for the optical communication applications (as described in Chap. 8).

(ii) The polarization independence
In the dispersion Eqs. (6.1) and (6.2), the maximum of the absolute value of the second term on the right-hand side is π. For the ultrahigh-order modes, the deletion of the second term will not introduce a significant error.

Consequently, both dispersion equations for TE and TM modes can be simply approximated by

$$\kappa_1 \mathrm{d} = m\pi. \tag{6.30}$$

It means that the ultrahigh-order modes are independent on the polarization of incident light.

(iii) Responding to a tiny change with a high sensitivity
Because the ultrahigh-order modes have a longer propagation time in the guiding layer, any tiny change of λ, n_1, and d will cause a dramatic variation of N. The theoretical sensitivity s is defined as the derivative of the effective RI with respect to a certain characteristic parameter, that is,

$$s = \frac{\mathrm{d}N}{\mathrm{d}\xi}, \tag{6.31}$$

where ξ represents λ, n_1, or d. By the total differential of Eq. (6.30), we obtain

$$\frac{\partial N}{\partial n_1} = \frac{n_1}{N}, \tag{6.32}$$

$$\frac{\partial N}{\partial \lambda} = \frac{n_1^2 - N^2}{N\lambda}, \tag{6.33}$$

$$\frac{\partial N}{\partial \mathrm{d}} = \frac{n_1^2 - N^2}{N\mathrm{d}}. \tag{6.34}$$

From the above three equations, it is clear that the sensitivity is in inverse proportion to the effective RI. As a result, the sensitivity of the ultrahigh-order modes ($N \to 0$) is large and this property is extremely useful for designing the optical sensors and modulators.

(iv) The slow light effect

According to Eq. (6.33), tiny change of wavelength can generate dramatic variation of the effective RI. It indicates that the ultrahigh-order modes have the strong dispersion property and the slow light effect. Using Eq. (6.30), we obtain the group velocity of the ultrahigh-order mode

$$v_g = \frac{\mathrm{d}\omega}{\mathrm{d}\beta} = \frac{N}{n_1} \cdot \frac{c}{(n_1 + \omega \mathrm{d}n_1/\mathrm{d}\omega)}, \tag{6.35}$$

where the right-hand side of the equation is composed of two contributions: One originates from the first-order dispersion of the material and the other from N/n_1, which is called as the slow light factor and is proportional to the effective RI of the ultrahigh-order modes. The deduced results offer us a new physical mechanism for realizing the wideband slow light, and more details will be given in the next subsection.

6.4 WideBand Slow Light

Slow light, emerging as a technology for optical delays, is expected to be instrumental in enabling various applications, e.g., optical buffer, on next-generation optical communication networks. According to [10], the ideal slow light devices are those whose RI and attenuation characteristics should be able to maximize the slow-down factor and minimize distortion by the frequency-dependent attenuation and dispersion. Generally, in the conventional slow light schemes [11–13], the group index (slow-down factor) is given by $n_g = n + \omega \frac{dn}{d\omega}$. Given that n_g is much larger than n, then

$$n_g \approx \omega \frac{\Delta n}{\Delta \omega}. \tag{6.36}$$

If Δn is a fixed value, the normalized bandwidth $\Delta n / \Delta \omega$ must be narrow for n_g to be extremely large. That is the reason why the reduction of the group velocity of light is always via exploiting the large dispersion associated with nearby optical resonances. However, it implies that there is a limitation of delay-bandwidth product (DBP), which refers to the fundamental trade-off between the total group delay Δt and the bandwidth $\Delta \omega$. In other words, the realization of the wideband slow light is quite difficult by using those conventional slow light schemes.

When the actual material dispersion is neglected, the group velocity of the ultrahigh-order modes (see Eq. (6.35)) can be simplified as follows:

$$v_g \approx \frac{Nc}{n_1^2}. \tag{6.37}$$

Unlike the conventional slow light schemes, the mechanism of ultrahigh-order modes assisted slow light [14, 15] does not rely on enlarging the dispersive term $\omega dn/d\omega$, but rather on tuning the effective refractive index N to zero. The underlying physics can be interpreted in terms of the anomalous dispersion of the ultrahigh-order modes, which is depicted in Fig. 6.9. It is demonstrated that anomalous dispersion curves of the SMCW structure exhibit an extremely flattened region (slow light region) in the vicinity of zero wave number ($N \rightarrow 0$). Consequently, this wideband slow light scheme is extremely simple and not subject to limitations of DBP. Another concerning issue is whether the ultrahigh-order modes could be efficiently confined in the guiding layer over a long transmission distance. This should be feasible, since the decay coefficient in the metal films can be written as $\alpha = k_0\sqrt{-\varepsilon_2} > 0$, which is quite large, as $|\varepsilon_2| \gg 1$ for a visible or red light beam. That means the ultrahigh-order modes are still confined better and its corresponding propagation loss is small even the existence of metal films.

The schematic diagram of the SMCW for verifying the wideband slow light effect is illustrated in Fig. 6.10. An additional silver stripe (about 500 nm thick and 1.1 cm wide) is sputtered on the middle of the coupling layer (sliver) to prevent

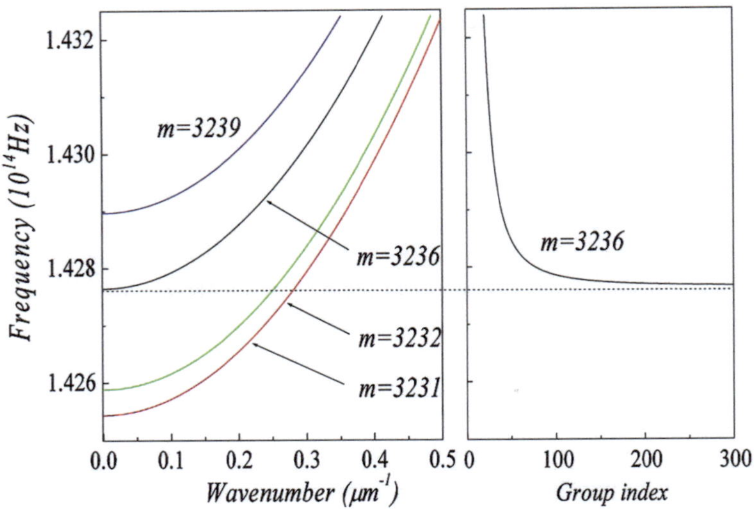

Fig. 6.9 Dispersion curves and group index characteristics of the ultrahigh-order modes

light leakage. Two photodiodes are set up to detect the light intensity. The first one (PD1) is placed in the reflected light beam, which serves as the reference beam. The second photodiode (PD2) and an oscilloscope are used to measure the time delay of the slow light beam. In [14], the light source is a collimated light beam and a group velocity less than 0.017c has been achieved. In [15], the light source is replaced by a femtosecond pulse, which has a wide frequency span over 10 THz (FWHM). The experimental measured time delay at different incident angles is illustrated in Fig. 6.11. Typical spectra of the input and output pulses are shown in the inset, which proves the capability of generating wideband slow light. The maximum DBP exceeding 1400 has been achieved.

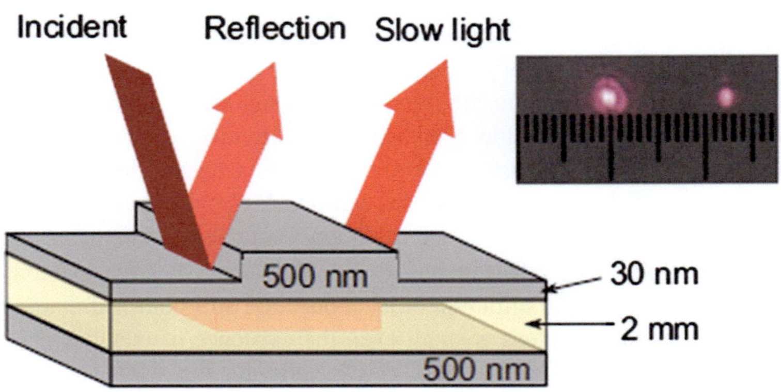

Fig. 6.10 Schematic layout of the SMCW for verifying the slow light effect

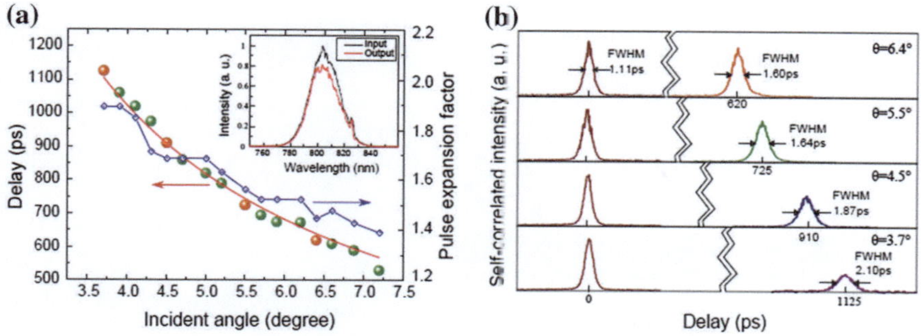

Fig. 6.11 **a** Measured absolute time delay and pulse expansion at different incident angles. Self-correlation width of the incident pulse is 1.11 ps. The solid curve is the theoretical fitting. (*Inset*) A typical input and output spectra. **b** Self-correlation trace of the optical pulse at each time delay denoted by the *yellow filled circles* plotted in (**a**)

6.5 Conical Reflection [16]

We start with a comparison between a conventional micron-scale dielectric waveguide and a SMCW, as shown in Fig. 6.12. From the perturbation theory [17, 18], the coupling coefficient between two modes by periodic perturbation can be written in the form of

$$\kappa_c = \frac{\kappa^2}{\beta^2 \Lambda \omega_{\mathrm{eff}}} \left| \int f(x) \mathrm{e}^{i2\Delta\beta x} \mathrm{d}x \right|, \tag{6.38}$$

where κ is the transverse component of the wave vector, Λ is the perturbation period, ω_{eff} is the effective thickness, $f(x)$ is the perturbation function along the propagation direction, and $\Delta\beta = (\beta_2 - \beta_1)/2$ is the propagation constant mismatch between the coupled modes. For a conventional micron-scale optical waveguide with scattering perturbation, the right integral term equals zero over a long distance. The coefficient $\kappa^2/\beta^2 = n^2/N^2 - 1$ of low-order modes is also rather small because their effective RI is smaller but very close to the RI of the guided layer, i.e., $N \approx n$. The result is that the inter- and intra-mode coupling is rather weak in uniform thin dielectric waveguides. Therefore, periodic perturbation would be necessary for efficient coupling of two different order guided modes over long propagating length. Usually, only short m lines (see Fig. 6.12b) can be observed because of mode coupling induced by the scattering in the guided layer.

On the other hand, the coefficient κ^2/β^2 of the ultrahigh-order modes is orders of magnitude larger, since their effective RI approaches to zero. Therefore, the mode coupling between ultrahigh-order modes is greatly enhanced. By denoting the scattering perturbation as a $\delta(\vec{r})$ function, where $\vec{r}$ is in the $x - y$ plane, then the coupling of ultrahigh-order modes can be recast as follows:

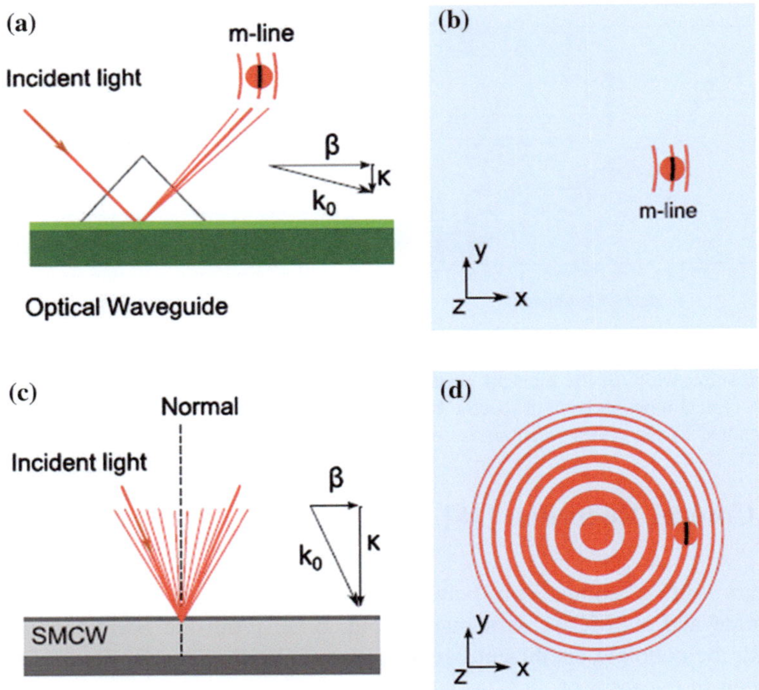

Fig. 6.12 Excitation of the m line in an all-dielectric waveguide and a SMCW. **a** Prism coupling of light into a thin dielectric optical waveguide. **b** A multiple short m line occurs when there is coupling among different guided modes due to light scattering. **c** Light coupling into SMCW by means of the free-space coupling technique. **d** The conical reflection pattern of SMCW, which is actually closed m-line circles

$$\kappa_c = \frac{\kappa^2}{\beta^2 \Lambda \omega_{\text{eff}}} \left| \int R\delta(\vec{r}) e^{i2\Delta\beta \vec{r}} \mathrm{d}\vec{r} \right| = \frac{\kappa^2 R}{\beta^2 \Lambda \omega_{\text{eff}}}, \tag{6.39}$$

where R is a defined parameter including total scattering perturbation and is of order unity. Equation (6.39) suggests that strong coupling among all mode orders and in all directions (see Fig. 6.12d) for the ultrahigh-order modes is possible without periodic perturbation. Consequently, the excited ultrahigh-order modes can easily be scattered to various directions of the same order and with other mode orders by the imperfection in the guiding layers, such as Rayleigh scattering. Due to the thin upper-layer cladding, leakage of the guided mode power cannot be neglected. The interesting result is that besides the normal reflection of light, there is also a conical reflection phenomenon at the certain critical angles.

The experimental setup is depicted in Fig. 6.13a. A beam from a He–Ne laser was slightly focused by a long focal length lens, reflected by the SMCW placed at the focus and then projected on a screen. There was a hole in the screen to let through the incident light. The pattern on the screen was recorded by use of a

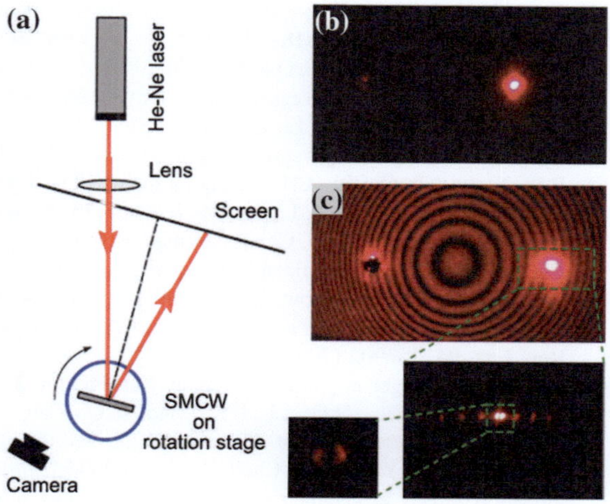

Fig. 6.13 Experimental setup for the observation of conical reflection by SMCW. **a** Top view of the setup. **b** Normal reflection away from the mode coupling condition. **c** The observed extraordinary conical reflection pattern

camera. The SMCW is composed of a 90 μm thick of glass slab sandwiched by two layers of silver film (34 nm upper layer and 200 nm substrate). Away from critical angles, the incident light experienced normal reflection, as shown in Fig. 6.13b. The left blur spot is the hole in the screen for laser incidence. The right blur spot is the reflection light projected on the screen because of scattering due to the imperfection of either the metal-cladding layer or the glass slab. When the incident angle satisfied the mode coupling condition, the extraordinary reflection (conical reflection) emerged as a concentric ring pattern projected on the screen, as shown in Fig. 6.13c. The intensity of each ring was measured to be equal in all directions, with outer rings having a slightly lower intensity. It should be mentioned that no microstructure was pre-introduced in the guided layer or on the cladding layer. Therefore, the ultrahigh-order mode assisted conical reflection is different from Newton's rings or multiple-beam interference as in Fabry–Perot resonance.

References

1. H. Shin, S. Fan, All-angle negative refraction for surface plasmon waves using a metal-dielectric-metal structure. Phys. Rev. Lett. **96**, 3907 (2006)
2. I.P. Kaminow, W.L. Mammer, H.P. Weber, Metal-clad optical waveguides: analytical and experimental study. Appl. Opt. **13**, 396 (1974)
3. F. Zhou, Z.Q. Cao, Y. Jiang, Q.S. Shen, Method to identify mode order in double metal-cladding waveguide. Acta Opt. Sin. **22**, 665 (2002). (in Chinese)
4. D. Marcuse, Coupled mode theory of round optical fibers. Bell Syst. Tech. J. **52**, 817 (1973)
5. P.K. Tien, R. Ulrich, R.J. Martin, Modes of propagating light waves in thin deposited semiconductor films. Appl. Phys. Lett. **14**, 291 (1969)
6. M.L. Dakss, L. Kuhn, P.F. Heidrich, B.A. Scott, Grating coupler for efficient excitation of optical guided waves in thin films. Appl. Phys. Lett. **16**, 523 (1970)

7. P.K. Tien, R.J. Martin, Experiments on light waves in a thin tapered film and a new light-wave coupler. Appl. Phys. Lett. **18**, 398 (1971)
8. H.G. Li, Z.Q. Cao, H.F. Lu, Q.S. Shen, Free-space coupling of a light beam into a symmetrical metal-cladding optical waveguide. Appl. Phys. Lett. **83**, 2757 (2003)
9. H.F. Lu, Z.Q. Cao, H.G. Li, Q.S. Shen, Study of ultrahigh-order modes in a symmetrical metal-cladding optical waveguide. Appl. Phys. Lett. **85**, 4579 (2004)
10. R.S. Tucker, P.-C. Ku, C.J. Chang-Hasnain, Slow-light optical buffers: capabilities and fundamental limitations. J. Lightwave Technol. **23**, 4046 (2005)
11. P.C. Ku, F. Sedgwick, C.J. Chang-Hasnain, P. Palinginis, T. Li, H. Wang, S.W. Chang, S.L. Chuang, Slow light in semiconductor quantum wells. Opt. Lett. **29**, 2291 (2004)
12. Y. Okawachi, M.S. Bigelow, J.E. Sharping, Z. Zhu, A. Schweinsberg, D.J. Gauthier, R.W. Boyd, A.L. Gaeta, Tunable all-optical delays via Brillouin slow light in an optical fiber. Phys. Rev. Lett. **94**, 153902 (2005)
13. T. Baba, Slow light in photonic crystals. Nat. Photon. **2**, 465 (2008)
14. W. Yuan, C. Yin, H.G. Li, P.P. Xiao, Z.Q. Cao, Wideband slow light assisted by ultrahigh-order modes. J. Opt. Soc. Am. B **28**, 968 (2011)
15. Y.L. Zheng, W. Yuan, X.F. Chen, Z.Q. Cao, Wideband slow-light modes for time delay of ultrashort pulses in symmetrical metal-cladding optical waveguide. Opt. Express **20**, 9409 (2012)
16. Y.L. Zheng, Z.Q. Cao, X.F. Chen, Conical reflection of light during free-space coupling into a symmetrical metal-cladding waveguide. J. Opt. Soc. Am. A: **30**, 1901 (2013)
17. A. Yariv, M. Nakamura, Periodic structures for integrated optics. IEEE J. Quantum Electron. **13**, 233 (1977)
18. K. Okamoto, *Fundamentals of Optical Waveguides* (Elsevier, Burlington, 2006)

Chapter 7
Goos–Hänchen Shift

Abstract This chapter introduces the readers to the related issues of the non-specular reflection effects and in particular of the Goos–Hänchen (GH) shift. We first briefly review two causality paradoxes in the optical reflection and their corresponding solutions with the consideration of GH time. Then, we elaborately describe the current work on the theoretical explanation and the experimental enhancement of the GH shift. Finally, we give a unified theory for all non-specular reflection effects.

Keywords Non-specular reflection · Causality paradox · Goos–Hänchen shift · Imbert–Fedorov shift · Group velocity · Stationary-phase approach · Enhancement

7.1 Obstacle in the Ray Theory

7.1.1 Contradiction Between the Ray Theory and the Electromagnetic Field Theory

It is well known that the reflected point will undergo a lateral discrepancy from its incident counterpart when a bounded light beam is totally reflected by an interface between two different materials (see Fig. 7.1). This discrepancy is called as the Goos–Hänchen (GH) shift [1], and the magnitude is given by [2]

$$\Delta z = \frac{2}{\kappa_1}\frac{\partial \phi}{\partial \theta}, \tag{7.1}$$

where $\kappa_1 = k_0 n_1 \cos\theta$ is the vertical component of the wave vector, $k_0 = 2\pi/\lambda$ is the wave number with light wavelength λ in the free space, θ is the incident angle, and -2ϕ is the total reflection phase shift between the reflected point B and the incident point A.

Is the above statement right? Let us consider the light ray propagated in the waveguide, which is shown in Fig. 7.2. From the ray theory described in Chap. 2,

X. Wang et al., *Progress in Planar Optical Waveguides*,
Springer Tracts in Modern Physics 266, DOI 10.1007/978-3-662-48984-0_7

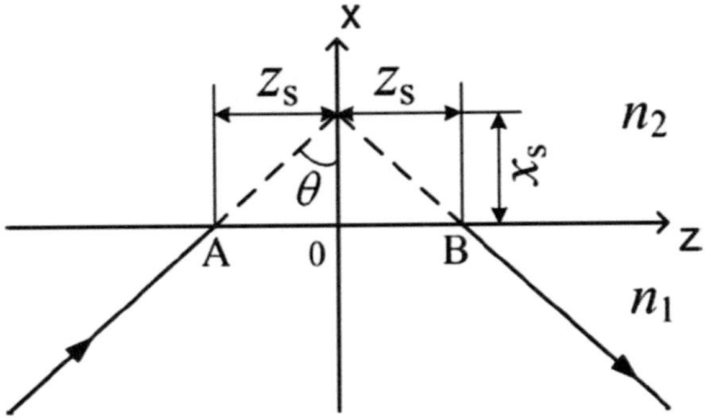

Fig. 7.1 GH shift at the total reflection between two semi-infinite medium

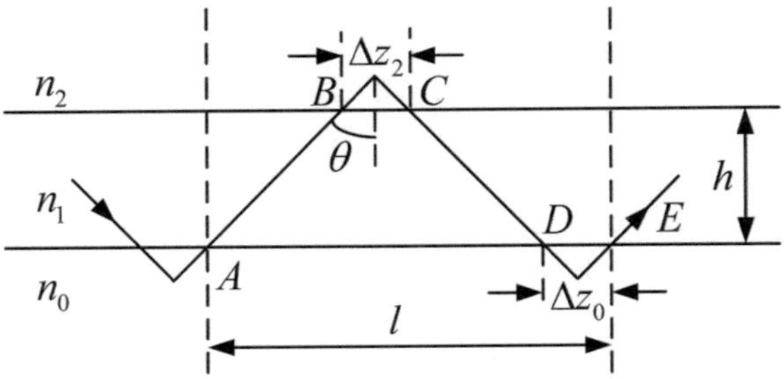

Fig. 7.2 The light ray propagated in the waveguide

we know that after the light ray traveling through one period (the area between two vertical dotted lines), the accumulated phase shift along the vertical direction of the incident plane must be an integer multiple of 2π, namely

$$\kappa_1 h = m\pi + \phi_{10} + \phi_{12}, \quad (m = 0, 1, 2, \cdots). \tag{7.2}$$

According to the stationary-phase approach [2], where $-2\phi_{10}$ is the phase shift of light ray reflected from point D to point E, and $-2\phi_{12}$ is the phase shift of light ray reflected from point B to point C. The ray theory predicts that the accumulated phase shift of one period (i.e., the light ray travels from point A to point B, then is reflected at the top interface to point C, sequentially travels to point D, and is finally reflected at the bottom interface to point E) is

$$\Phi_1 = 2k_0 n_1 \frac{h}{\cos\theta} - 2(\phi_{10} + \phi_{12}). \tag{7.3}$$

In the electromagnetic field theory, the accumulated phase shift of one guided mode with propagation constant β traveling along the distance l is

$$\Phi_2 = \beta l. \tag{7.4}$$

If the ray theory is self-consistent with the electromagnetic field theory, we get

$$\Phi_1 - \Phi_2 = 2m\pi. \tag{7.5}$$

Combined with Eq. (7.2), Φ_1 can be rewritten as

$$\Phi_1 = 2k_0 n_1 \frac{h}{\cos\theta} - 2(k_0 n_1 \cos\theta \cdot h - m\pi) = 2\beta h \tan\theta + 2m\pi. \tag{7.6}$$

The length of one period is

$$l = 2h\tan\theta + (\Delta z_0 + \Delta z_2), \tag{7.7}$$

and it is easy to obtain

$$\Phi_1 - \Phi_2 = 2m\pi - \beta(\Delta z_0 + \Delta z_2). \tag{7.8}$$

Apparently, two theories are not self-consistent with each other.

7.1.2 The Addition of Lateral Phase Shift

What cause the above contradiction between the ray theory and the electromagnetic field theory? Resch et al. [3] proposed that the total phase shift at the total reflection should contain two components: the vertical phase shift component and a lateral phase shift component. The total phase shift is written as

$$\Phi_{TOT} = -2\phi + k_0 n_1 \sin\theta \cdot \Delta z. \tag{7.9}$$

Based on the above assumption, Eq. (7.3) is wrong and should be modified as

$$\begin{aligned} \Phi_1 &= 2k_0 n_1 \frac{h}{\cos\theta} - 2(\phi_{10} + \phi_{12}) + \beta(\Delta z_0 + \Delta z_2) \\ &= 2\beta h \tan\theta + \beta(\Delta z_0 + \Delta z_2) + 2m\pi. \end{aligned} \tag{7.10}$$

Now, it is easy to verify that Eq. (7.5) is correct. By adding the lateral phase shift component to the total phase shift, the contradiction between two theories has been solved. The above equation can also be expressed as

$$\Phi_1 = 2\beta h_{\text{eff}} \tan\theta + 2m\pi, \tag{7.11}$$

where h_{eff} is the effective thickness of the waveguide. Equation (7.11) indicates that the propagation behavior of the light ray in the waveguide is the same as that in a thin film with a thickness of h_{eff} by omitting the phase shift of the total reflection.

7.2 Causality Paradoxes in Gires–Tournois Interferometer and Plasma Mirror

7.2.1 Causality Paradox in Gires–Tournois Interferometer [4]

Let us consider the configuration of the Gires–Tournois interferometer shown in Fig. 7.3 and set $n_1 > n_3 > n_2$. When the incident angle is larger than the critical angle of the total reflection between medium n_1 and medium n_3, the electromagnetic fields in the medium n_2 and medium n_3 are evanescent. The reflection coefficient is

$$r = \frac{B_1}{A_1} = \frac{\exp(-i2\phi_{12}) + r_{23}\exp(-2\alpha_2 \mathrm{d})}{1 + r_{23}\exp(-i2\phi_{12})\exp(-2\alpha_2 \mathrm{d})}, \tag{7.12}$$

where

$$\phi_{12} = \begin{cases} \arctan\left(\dfrac{\alpha_2}{\kappa_1}\right), & \text{for TE mode,} \\ \arctan\left(\dfrac{n_1^2}{n_2^2}\dfrac{\alpha_2}{\kappa_1}\right), & \text{for TM mode,} \end{cases} \tag{7.13}$$

$$r_{23} = \begin{cases} \dfrac{\alpha_2 + \alpha_3}{\alpha_2 - \alpha_3}, & \text{for TE mode,} \\ \dfrac{n_3^2\alpha_2 + n_2^2\alpha_3}{n_3^2\alpha_2 - n_2^2\alpha_3}, & \text{for TM mode,} \end{cases} \tag{7.14}$$

$$\begin{cases} \kappa_1 = k_0 n_1 \cos\theta, \\ \alpha_2 = k_0\left(n_1^2\sin^2\theta - n_2^2\right)^{1/2}, \\ \alpha_2 = k_0\left(n_1^2\sin^2\theta - n_3^2\right)^{1/2}. \end{cases} \tag{7.15}$$

If all mediums are non-absorbing and non-dispersive, the relationship between two complex amplitudes in the medium n_1 can be expressed as

$$B_1 = A_1\exp(-i2\phi), \tag{7.16}$$

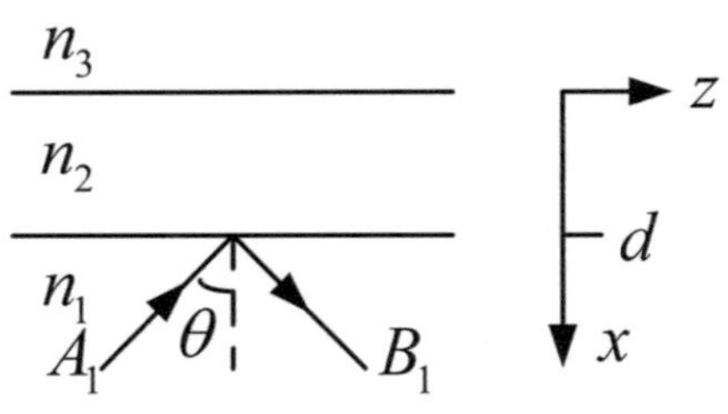

Fig. 7.3 The Gires–Tournois interferometer

where -2ϕ is the phase shift of the total reflection from the Gires–Tournois interferometer. On comparing Eq. (7.12) with Eq. (7.16), we find

$$\tan\phi = \tan\phi_{12}\frac{1 - r_{23}\exp(-2\alpha_2 \mathrm{d})}{1 + r_{23}\exp(-2\alpha_2 \mathrm{d})}. \tag{7.17}$$

The group delay time is written as

$$t_g = -2\frac{\partial\phi}{\partial\omega} = -\frac{(8\alpha_2 \mathrm{d}/\omega)\tan\phi_{12}\cdot r_{23}\exp(-2\alpha_2 \mathrm{d})}{[1 + r_{23}\exp(-2\alpha_2 \mathrm{d})]^2 + \tan^2\phi_{12}\cdot[1 - r_{23}\exp(-2\alpha_2 \mathrm{d})]^2}. \tag{7.18}$$

Owing to $n_3 > n_2$, therefore, α_2 is larger than α_3, and the Fresnel coefficient r_{23} is always larger than zero for both two linear polarizations. It can be concluded that the group delay time in Eq. (7.18) is negative, i.e., $t_g < 0$. However, the negative group delay time violates the relativistic causality. Tournois [4] cannot interpret this unbelievable result, which is termed as a causality paradox.

7.2.2 *Interpretation of Causality Paradox in the Gires–Tournois Interferometer*

Actually, the encountered problem in the above causality paradox is identical with that in the waveguide. The solution to the causality paradox is also that the lateral phase shift component should be added. The total phase shift of the total reflection from the Gires–Tournois interferometer is determined by Eq. (7.9), and the total delay time should be

$$\frac{\partial\Phi_{\mathrm{TOT}}}{\partial\omega} = -2\frac{\partial\phi}{\partial\omega} + \frac{n_1}{c}\sin\theta\cdot(\Delta z), \tag{7.19}$$

where the first term is the group delay time t_g, which is resulted from the dispersion of the phase shift, and the second term is the GH time, which is originated from the lateral phase shift and can be derived from Eq. (7.1)

$$t_{\mathrm{GH}} = \frac{n_1}{c}\sin\theta\cdot\left(\frac{2}{k_0 n_1\cos\theta}\cdot\frac{\partial\phi}{\partial\theta}\right) = \frac{2\tan\theta}{\omega}\frac{\partial\phi}{\partial\theta}. \tag{7.20}$$

With the addition of GH time, Resch has proved that the total delay time of the total reflection is positive

$$t = t_g + t_{\mathrm{GH}} > 0, \tag{7.21}$$

namely the above causality paradox does not exist. The detailed derivation process of Eq. (7.21) is described in [3].

7.2.3 Causality Paradox Associated with Total Reflection upon the Plasma Mirror

Even though Resch had proved that the total delay time of total reflection is positive, Tournois did not accept Resch's opinion and put forward an example of the total reflection of a TM plane wave from vacuum upon an ideal non-absorbing plasma mirror to illustrate that the total delay time is negative in spite of taking the GH time into account [5].

The plasma frequency and the light frequency are denoted by ω_p and ω, respectively, and their ratio is marked as $u = \omega/\omega_p$. The RI of plasma is $n_p = -i(1 - u^2)^{1/2}/u$, where $0 < u < 1$. If a TM plane wave incidents upon the interface between the vacuum ($n_1 = 1$) and the plasma mirror with a incident angle θ, the phase shift -2ϕ can be derived from Eq. (7.13) and expressed as

$$\tan\phi = -\frac{u(1 - u^2\cos^2\theta)^{1/2}}{(1 - u^2)\cos\theta}. \tag{7.22}$$

The group delay time is

$$t_g = -2\frac{\partial\phi}{\partial\omega} = \frac{2\cos\theta}{\omega_p(1 - u^2\cos^2\theta)^{1/2}}\cdot\frac{(1 - u^2\cos 2\theta)}{(\cos^2\theta - u^2\cos 2\theta)}, \tag{7.23}$$

and it is clear that $t_g > 0$. The expression of GH time is

$$t_{\mathrm{GH}} = \frac{2\tan\theta}{\omega}\left(\frac{\partial\phi}{\partial\theta}\right) = -\frac{2\tan\theta\sin\theta}{\omega_p(1 - u^2\cos^2\theta)^{1/2}}\cdot\frac{(1 - u^2)}{(\cos^2\theta - u^2\cos 2\theta)}. \tag{7.24}$$

It is easy to confirm that $t_{\mathrm{GH}} < 0$.

When the incident angle θ satisfies

$$u^2 < \frac{\tan^2\theta - 1}{\tan^2\theta - \cos 2\theta}, \tag{7.25}$$

the total delay time is

$$t = t_g + t_{\mathrm{GH}} < 0. \tag{7.26}$$

The group delay time t_g, GH time t_{GH}, and the total delay time ($t = t_g + t_{\mathrm{GH}}$) as functions of u are depicted in Fig. 7.4. It is easy to see that when $u < 0.8$, the total delay time t is negative. The example of the total reflection upon plasma mirror illustrates that the causality paradox does not exist since the group delay time t_g is positive. However, by following the Resch's opinion, namely adding the GH time t_{GH} to the group delay time, the total delay time is still possible to be negative.

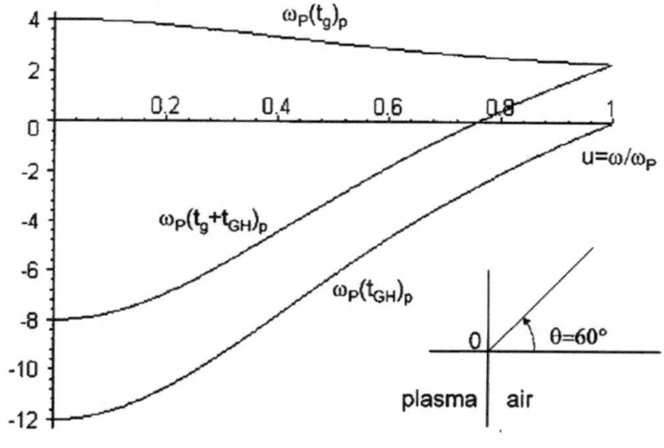

Fig. 7.4 The group delay time, GH time, and the total delay time as functions of u

Therefore, Tournois concluded that Resch's opinion is wrong and there is only the existence of the phase shift-induced group delay time and no existence of the GH time.

7.2.4 *Interpretation of Causality Paradox in the Plasma Mirror [6]*

In our opinion, the GH time associated with the total reflection does always exist, but its expression written by Eq. (7.24) is incomplete. Actually, the reflection point of a TM plane wave totally reflected upon the plasma mirror is not located at the interface but at the front of interface (see Fig. 7.5) since $n_p^2 < 0$. By denoting the reflection point as O and the lateral shift AB as Δz, the total reflection phase shift will consist of three parts: one part is the conventional reflection phase shift -2ϕ, the second part is the lateral shift-induced phase shift, and the third part is the accumulated phase shift of the light ray propagating from point O to point A and from point B to point O, namely

$$\Phi_{\mathrm{TOT}} = -2\phi + k_0 n_1(\Delta z)(\sin\theta - 1/\sin\theta), \tag{7.27}$$

Therefore, the accurate expression for GH time is

$$\tau_{\mathrm{GH}} = \frac{n_1}{c}(\Delta z)(\sin\theta - 1/\sin\theta) = \frac{2(1-u^2)\cos\theta}{\omega_p(1-u^2\cos^2\theta)^{1/2}(\cos^2\theta - u^2\cos 2\theta)}. \tag{7.28}$$

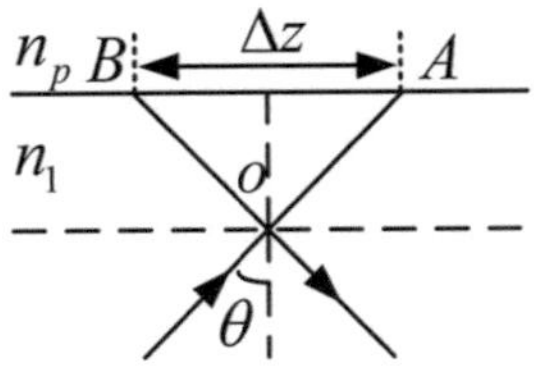

Fig. 7.5 Total reflection upon plasma mirror

Apparently, $\tau_{GH} > 0$. The total delay time is also positive since $t_g > 0$. Thus, one can claim that the causality paradox in the plasma mirror does not exist.

7.2.5 Detailed Analysis of the Optical Waveguide

In order to test whether the opinion in above subsection is right or not, let us consider the symmetrical planar waveguide shown in Fig. 7.6. The guiding layer (vacuum) is symmetrically covered by the non-absorbing plasma. The dispersion equation for TM mode is

$$\kappa_1 h = m\pi + 2\phi, \quad (m = 0, 1, 2, \cdots), \tag{7.29}$$

where the phase shift -2ϕ of the total reflection at the upper interface and the bottom interface is defined by

$$\tan\phi = \frac{1}{n_p^2}\left(\frac{N^2 - n_p^2}{1 - N^2}\right)^{1/2}, \tag{7.30}$$

and N is the effective RI of the guided mode.The group velocity v_g of the guided mode is

$$\frac{1}{v_g} = \frac{\partial\beta}{\partial\omega} = \frac{N}{c} + k_0\frac{\partial N}{\partial\omega}, \tag{7.31}$$

Fig. 7.6 Ray model of the symmetrical plasma cladding planar waveguide

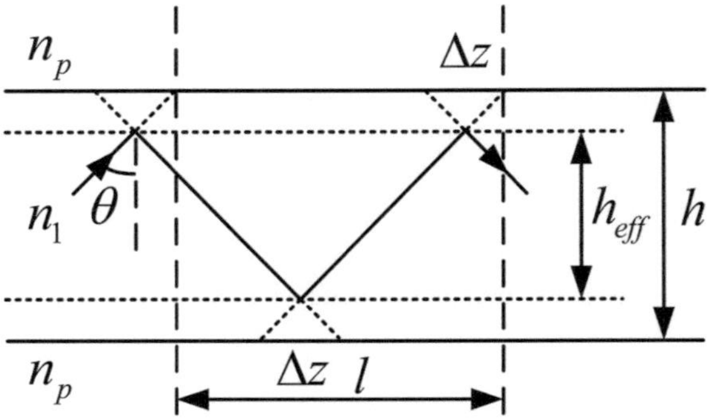

Applying the differential operation on Eq. (7.29), we obtain

$$\frac{\partial N}{\partial \omega} = \frac{1-N^2}{\omega N h_{\mathrm{eff}}}\left[h + 2\frac{1-n_p^2}{n_p^4(1-N^2)+\left(N^2-n_p^2\right)}\cdot\frac{2N^2-n_p^2}{k_0\left(N^2-n_p^2\right)^{1/2}}\right]. \tag{7.32}$$

By substituting Eq. (7.32) into Eq. (7.31), we get

$$\frac{1}{v_g} = \frac{N}{c} + \frac{1-N^2}{cNh_{\mathrm{eff}}}\cdot\left[h + 2\frac{1-n_p^2}{n_p^4(1-N^2)+\left(N^2-n_p^2\right)}\cdot\frac{2N^2-n_p^2}{k_0\left(N^2-n_p^2\right)^{1/2}}\right]. \tag{7.33}$$

Since length of one period for the light ray propagating in the waveguide is

$$l = 2h\tan\theta + 2(\Delta z) = 2h_{\mathrm{eff}}\tan\theta, \tag{7.34}$$

as a result, the propagation time of the guided mode traveling through one period l is

$$\begin{aligned}\tau_{\mathrm{total}} &= 2h_{\mathrm{eff}}\tan\theta/v_g\\ &= \frac{2\tan\theta}{cN}\left[h + 2\frac{n_p^2\left(1-n_p^2\right)}{n_p^4(1-N^2)+\left(N^2-n_p^2\right)}\cdot\frac{N^2}{k_0\left(N^2-n_p^2\right)^{1/2}}\right.\\ &\quad\left. + 2\frac{\left(1-n_p^2\right)\left(2N^2-n_p^2\right)}{n_p^4(1-N^2)+\left(N^2-n_p^2\right)}\cdot\frac{1-N^2}{k_0\left(N^2-n_p^2\right)^{1/2}}\right].\end{aligned} \tag{7.35}$$

Combined with $N = \sin\theta$, Eqs. (7.23 and 7.24), Eq. (7.35) can be recast as

$$\begin{aligned}\tau_{\mathrm{total}} &= \frac{2h}{c\cos\theta} - \frac{4\tan\theta\sin\theta}{\omega_p(1-u^2\cos^2\theta)^{1/2}}\cdot\frac{(1-u^2)}{(\cos^2\theta - u^2\cos 2\theta)}\\ &\quad + \frac{4\cos\theta}{\omega_p(1-u^2\cos^2\theta)^{1/2}}\cdot\frac{(1-u^2\cos 2\theta)}{(\cos^2\theta-u^2\cos 2\theta)}\\ &= \frac{2h}{c\cos\theta} + 2t_{\mathrm{GH}} + 2t_g.\end{aligned} \tag{7.36}$$

It is very clear that the total delay time τ_{total} consists of three parts, (i) the propagation time 2 $h/(c\cos\theta)$ of the light ray in the guiding layer, (ii) the GH time $2t_{\mathrm{GH}}$ at the two interfaces, and (iii) the group delay time $2t_g$ at the two interfaces. Note that h in Eq. (7.36) is the thickness of the guiding layer, and the GH time only contains the time of light ray propagating through the lateral shift Δz. To take account of the actual reflection points, we can rewrite Eq. (7.36) into

$$\tau_{\text{total}} = \frac{2h_{\text{eff}}}{c\cos\theta} + \frac{4\cos\theta}{\omega_p(1-u^2\cos^2\theta)^{1/2}} \cdot \frac{(1-u^2)}{(\cos^2\theta - u^2\cos 2\theta)} + \frac{4\cos\theta}{\omega_p(1-u^2\cos^2\theta)^{1/2}} \cdot \frac{(1-u^2\cos 2\theta)}{(\cos^2\theta - u^2\cos 2\theta)}. \tag{7.37}$$

where the first term is the propagation time of light ray in the effective thickness of the guiding layer, the second term is the actual GH time, which is defined by Eq. (7.28) and contains the time of light ray propagating through the lateral shift and the time of light ray propagating from the actual reflection point to the surface of plasma mirror, and the third term is still the group delay time. The conversion from Eq. (7.36) into Eq. (7.37) illustrates that the expression of GH time defined by Eq. (7.28) is correct.

In conclusion, the total delay time of the total reflection not only in the Gires–Tournois interferometer but also in the plasma mirror is always positive because, besides the group delay time induced by the reflection phase shift, the total delay time also contains the GH time. Therefore, both above causality paradoxes do not exist.

7.3 Generalized Form of the GH Time [7]

We have demonstrated the existence of GH time in the waveguide constituted by the non-absorbing plasma mirror. Here, the generalized planar waveguide, whose each layer is dispersive, is discussed in detail to study the group delay time and GH time of the total reflection and to derive a generalized form of the GH time.

7.3.1 Group Velocity of the Planar Waveguide

Let us consider a planar waveguide shown in Fig. 7.7. The material of each layer is dispersive, and the dispersion equation of the waveguide is

$$2\frac{\omega}{c}n_1 h\cos\theta = 2m\pi + 2\phi_{12} + 2\phi_{13}, \quad (m = 0, 1, 2, \cdots), \tag{7.38}$$

$$\tan\phi_{1j} = \begin{cases} \dfrac{\left(n_1^2\sin^2\theta - n_j^2\right)^{1/2}}{n_1\cos\theta}, & \text{For TE mode,} \\[2ex] \dfrac{n_1^2}{n_j^2}\cdot\dfrac{\left(n_1^2\sin^2\theta - n_j^2\right)^{1/2}}{n_1\cos\theta}, & \text{For TM mode,} \end{cases} \tag{7.39}$$

where ω is the angular frequency, c is the light velocity in the vacuum, h is the thickness of the guiding layer, θ is the incident angle, m is the order of the guided mode, and $j = 2$, 3 represent the medium of two cladding layers, respectively.

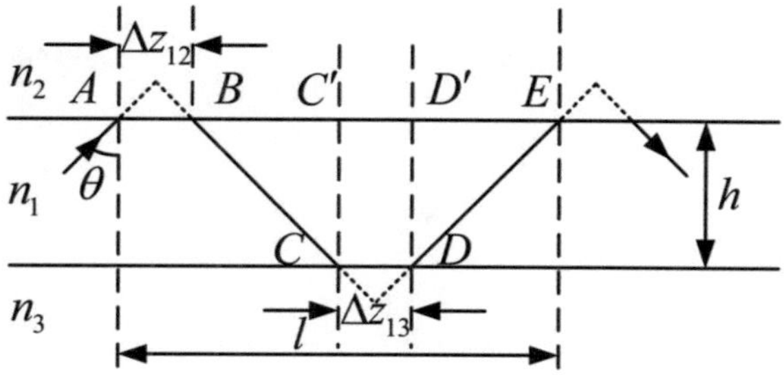

Fig. 7.7 A generalized planar waveguide and its ray model

The group RI m_1 is defined as

$$m_1 = n_1 + \omega dn_1/d\omega. \tag{7.40}$$

Taking the derivation of Eq. (7.40) with respect to the angular frequency, we obtain

$$\begin{aligned} &2\frac{m_1 h}{c}\cos\theta - 2\frac{\omega}{c}n_1 h\sin\theta\cdot\frac{\mathrm{d}\theta}{\mathrm{d}\omega} - 2\frac{\partial\phi_{12}}{\partial n_1}\frac{\mathrm{d}n_1}{\mathrm{d}\omega} - 2\frac{\partial\phi_{12}}{\partial n_2}\frac{\mathrm{d}n_2}{\mathrm{d}\omega} - 2\frac{\partial\phi_{12}}{\partial\theta}\frac{\mathrm{d}\theta}{\mathrm{d}\omega} \\ &- 2\frac{\partial\phi_{13}}{\partial n_1}\frac{\mathrm{d}n_1}{\mathrm{d}\omega} - 2\frac{\partial\phi_{13}}{\partial n_2}\frac{\mathrm{d}n_2}{\mathrm{d}\omega} - 2\frac{\partial\phi_{13}}{\partial\theta}\frac{\mathrm{d}\theta}{\mathrm{d}\omega} = 0. \end{aligned} \tag{7.41}$$

According to the definition of the group delay $\tau_g = -2\partial\phi/\partial\omega$, we get

$$\begin{aligned} \tau_g^{(1,2)} &= -2\frac{\partial\phi_{12}}{\partial\omega} = -2\frac{\phi_{12}}{\partial n_1}\frac{\mathrm{d}n_1}{\mathrm{d}\omega} - 2\frac{\partial\phi_{12}}{\partial n_2}\frac{\mathrm{d}n_2}{\mathrm{d}\omega}, \\ \tau_g^{(1,3)} &= -2\frac{\partial\phi_{13}}{\partial\omega} = -2\frac{\partial\phi_{13}}{\partial n_1}\frac{\mathrm{d}n_1}{\mathrm{d}\omega} - 2\frac{\partial\phi_{13}}{\partial n_2}\frac{\mathrm{d}n_2}{\mathrm{d}\omega}. \end{aligned} \tag{7.42}$$

By substituting Eq. (7.42) into Eq. (7.41), we find

$$\frac{\mathrm{d}\theta}{\mathrm{d}\omega} = \frac{2\frac{m_1 h}{c}\cos\theta + \tau_g^{(1,2)} + \tau_g^{(1,3)}}{2\frac{\omega}{c}n_1 h\sin\theta + 2\frac{\partial\phi_{12}}{\partial\theta} + 2\frac{\partial\phi_{13}}{\partial\theta}}. \tag{7.43}$$

The group velocity of the guided mode is expressed as

$$v_g = \frac{\mathrm{d}\omega}{\mathrm{d}\beta} = \frac{1}{\mathrm{d}\beta/\mathrm{d}\omega}, \tag{7.44}$$

where the propagation constant is

$$\beta = \frac{\omega}{c}n_1\sin\theta. \tag{7.45}$$

The derivative of the propagation constant with respect to the angular frequency is

$$\frac{\mathrm{d}\beta}{\mathrm{d}\omega}=\frac{m_1\sin\theta}{c}+\frac{\omega}{c}n_1\cos\theta\cdot\frac{\mathrm{d}\theta}{\mathrm{d}\omega}. \tag{7.46}$$

By substituting Eq. (7.43) and Eq. (7.46) into Eq. (7.44), the group velocity of the guided mode is

$$v_g=\frac{2h\tan\theta+\dfrac{2c}{\omega n_1\cos\theta}\dfrac{\partial\phi_{12}}{\partial\theta}+\dfrac{2c}{\omega n_1\cos\theta}\dfrac{\partial\phi_{13}}{\partial\theta}}{\dfrac{2m_1h}{c\cos\theta}+2\dfrac{m_1\tan\theta}{\omega n_1}\dfrac{\partial\phi_{12}}{\partial\theta}+2\dfrac{m_1\tan\theta}{\omega n_1}\dfrac{\partial\phi_{13}}{\partial\theta}+\tau_g^{(1,2)}+\tau_g^{(1,3)}}. \tag{7.47}$$

7.3.2 Generalized Form of the GH Time

The numerator in Eq. (7.47) is divided into three terms, and the physical meanings of the each term are

$$\begin{aligned}
&2h\tan\theta=BC'+D'E,\\
&\frac{2c}{\omega n_1\cos\theta}\cdot\frac{\partial\phi_{12}}{\partial\theta}=\Delta z_{12},\\
&\frac{2c}{\omega n_1\cos\theta}\cdot\frac{\partial\phi_{13}}{\partial\theta}=\Delta z_{13}.
\end{aligned} \tag{7.48}$$

Therefore, the sum of above three terms is the length l (see Fig. 7.7), which is the one period of the light ray propagating in the waveguide. The denominator in Eq. (7.47) consists of five terms. The physical meaning of the first term $2m_1h/(c\cos\theta)$ is the sum of propagating time of the light ray through BC and DE because $2h/\cos\theta$ is the total length of propagation paths BC and DE. $\tau_g^{(1,2)}$ and $\tau_g^{(1,3)}$ are the group delay times of the total reflection occurring at the interfaces AB and CD, respectively. The second and the third terms are expressed as

$$\begin{cases}
2\dfrac{m_1\tan\theta}{\omega n_1}\cdot\dfrac{\partial\phi_{12}}{\partial\theta}=\dfrac{m_1\sin\theta}{c}\Delta z_{12},\\
2\dfrac{m_1\tan\theta}{\omega n_1}\cdot\dfrac{\partial\phi_{13}}{\partial\theta}=\dfrac{m_1\sin\theta}{c}\Delta z_{13}.
\end{cases} \tag{7.49}$$

Apparently, the expression of Eq. (7.49) is quite different from the above definition of the GH time. Therefore, it is necessary to restudy the GH time from the aspect of its physical meaning. The total phase shift is

$$\phi_{\mathrm{TOT}}=-2\phi+k_z\Delta z=-2\phi+\frac{n_1\omega}{c}\sin\theta\cdot\Delta z. \tag{7.50}$$

By taking the derivation of total phase shift with respect to the angular frequency, we obtain

$$\frac{\partial \phi_{\mathrm{TOT}}}{\partial \omega} = -2\frac{\partial \phi}{\partial \omega} + \frac{n_1}{c}\sin\theta \cdot \Delta z + \frac{\omega}{c}\frac{\mathrm{d}n_1}{\mathrm{d}\omega}\sin\theta \cdot \Delta z = -2\frac{\partial \phi}{\partial \omega} + \frac{m_1}{c}\sin\theta \cdot \Delta z, \tag{7.51}$$

where the first term is still the group delay time, and the second term is the generalized form of the GH time, which is identical to Eq. (7.49) derived from the group velocity of waveguide. Therefore, in case of incidence medium being a dispersive material, the generalized form of the GH time is

$$\tau_{\mathrm{GH}} = \frac{m_1 \sin\theta}{c} \cdot \Delta z = 2\frac{m_1 \tan\theta}{\omega n_1} \cdot \frac{\partial \phi}{\partial \theta}. \tag{7.52}$$

A full analysis for the physical meaning of each term in the group velocity of Eq. (7.47) has been carried out. The expression of group velocity can be rewritten as

$$v_g = \frac{2h\tan\theta + \Delta z_{12} + \Delta z_{13}}{\frac{2m_1 h}{c\cos\theta} + \tau_{\mathrm{GH}}^{(1,2)} + \tau_g^{(1,2)} + \tau_{\mathrm{GH}}^{(1,3)} + \tau_g^{(1,3)}}. \tag{7.53}$$

It is clear that the propagation time of the length $2h\tan\theta$ is $2m_1h/(c\cos\theta)$, the propagation time of Δz_{12} is the sum of $\tau_{\mathrm{GH}}^{(1,2)}$ and $\tau_g^{(1,2)}$, and the propagation time of Δz_{13} is the sum of $\tau_{\mathrm{GH}}^{(1,3)}$ and $\tau_g^{(1,3)}$. Equation (7.53) evidently indicates that the total delay time of the total reflection is the sum of the group delay time and the GH time.

7.4 Theoretical Models for the GH Shift

7.4.1 Stationary-Phase Approach [2]

The reason for the emergence of GH shift is that the actual incident light is not an ideal monochromatic plane wave but a beam with a certain spatial spectral width. Consequently, the incident light can be decomposed into a series of monochromatic plane waves, and the wave vector of each one plane wave possesses a slightly different tangential component when compared to the other ones. In the total reflection, the reflection phase shift of each one plane wave is slightly different from that of the other ones and the actual reflected light beam is composed of a series of such reflected plane waves. As a result, there is a lateral shift, which is called as GH shift, between the maximal intensity position of the reflected light and that of the incident light.

Let us consider the interface between two semi-infinite medium shown in Fig. 7.1. When the incident angle is larger than the critical angle, namely

$\theta > \theta_c = \arcsin(n_2/n_1)$, where n_1 and n_2 are the RI of the two mediums and $n_1 > n_2$, the incident light will be totally reflected and its reflection coefficient is $r = \exp(i\varphi) = \exp(-i2\phi)$, where the reflection phase shift for the TM and the TE polarizations is

$$\phi_{\mathrm{TE}} = \arctan\left[\frac{\sqrt{n_1^2 \sin^2\theta - n_2^2}}{n_1 \cos\theta}\right], \tag{7.54}$$

$$\phi_{\mathrm{TM}} = \arctan\left[\left(\frac{n_1}{n_2}\right)^2 \frac{\sqrt{n_1^2 \sin^2\theta - n_2^2}}{n_1 \cos\theta}\right]. \tag{7.55}$$

To determine the magnitude of the lateral shift $2z_s$, let us consider a simple wave packet consisting of two plane waves with slightly different incident angles. Supposing that the z-component of the wave vector is $\beta \pm \Delta\beta$, the complex amplitude of the incident wave packet at the interface can be expressed as

$$A(z) = [\exp(i\Delta\beta z) + \exp(-i\Delta\beta z)]\exp(i\beta z) = 2\cos(\Delta\beta z)\exp(i\beta z). \tag{7.56}$$

$\Delta\beta$ is a small quantity, and the total reflection phase shift can be expanded by the differential formula

$$\phi(\beta \pm \Delta\beta) = \phi(\beta) \pm \frac{\mathrm{d}\phi}{\mathrm{d}\beta}\Delta\beta. \tag{7.57}$$

Therefore, the complex amplitude of the reflected wave packet at the interface is

$$\begin{aligned} B(z) &= \{\exp[i(\Delta\beta z - 2\Delta\phi)] + \exp[-i(\Delta\beta z - 2\Delta\phi)]\}\exp(i\beta z - 2\phi) \\ &= 2\cos[\Delta\beta(z - 2z_s)]\exp[i(\beta z - 2\phi)], \end{aligned} \tag{7.58}$$

where

$$z_s = \frac{\mathrm{d}\phi}{\mathrm{d}\beta}, \tag{7.59}$$

is the simple form for the lateral shift of the wave packet. The vertical distance between the actual reflected light beam and the theoretical reflected light beam predicted by the geometrical optics is

$$S = 2z_s\cos\theta = \frac{2\mathrm{d}\phi}{\mathrm{d}\beta}\cos\theta = \frac{2}{k_0 n_1}\frac{\mathrm{d}\phi}{\mathrm{d}\beta}. \tag{7.60}$$

The obtained expression of the GH shift is identical to Eq. (7.1), and its derivation procession is called as the stationary-phase approach, which is proposed by

Artmann [2]. Substituting Eqs. (7.54) and (7.55) into Eq. (7.60), the GH shift at the total reflection between two semi-infinite mediums is obtained as

$$S_{\mathrm{TE}} = \frac{2\sin\theta}{k_0\sqrt{n_1^2\sin^2\theta - n_2^2}}, \tag{7.61}$$

$$S_{\mathrm{TM}} = \frac{S_{\mathrm{TE}}}{\left[(n_1/n_2)^2 + 1\right]\sin^2\theta - 1}. \tag{7.62}$$

7.4.2 *Gaussian Beam Model [8]*

As shown in Fig. 7.8, an interface is illuminated by a Gaussian light, whose field at the interface $z = 0$ is expressed as

$$\psi_i(x, z = 0) = \exp\left(-\frac{x^2}{2w_x^2} + i\beta_0 x\right). \tag{7.63}$$

This Gaussian shape beam can be rewritten as the Fourier integral

$$\psi_i(x, z = 0) = \frac{1}{\sqrt{2\pi}}\int A(\beta)\exp(i\beta x)\mathrm{d}\beta, \tag{7.64}$$

where $w_x = w_0 \sec\theta$ with w_0 as the width of waist, and β_0 is the x-component of the wave vector with the incident angle θ. The angular spectrum distribution of the incident beam is

$$A(\beta) = w_x \exp\left[-(w_x^2/2)(\beta - \beta_0)^2\right]. \tag{7.65}$$

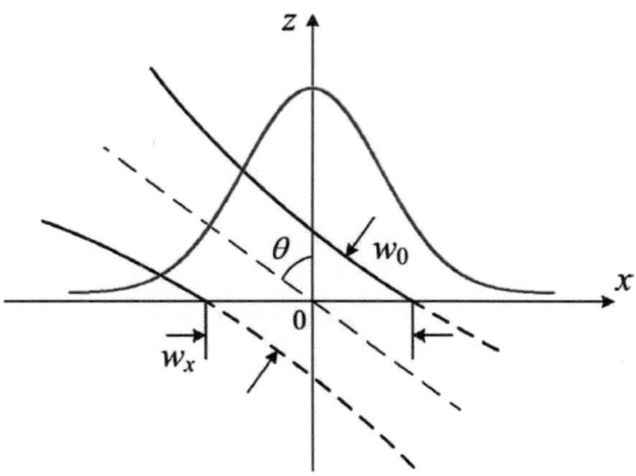

Fig. 7.8 The Gaussian beam model

Upon contacting with the interface, the incident beam is reflected and the reflection coefficient of each frequency component is non-uniform. The profile of the reflected beam is given by

$$\psi_r(x, z = 0) = \frac{1}{\sqrt{2\pi}} \int r(\beta) A(\beta) \exp(i\beta x) \mathrm{d}\beta. \tag{7.66}$$

The corresponding GH shift can be calculated from the above integral by finding the x-component position of max $|\psi_r(x, z = 0)|$. The range of integral is $(-k, k)$, where k is the wave vector in the incident medium, because the range of incident angle is $(-\pi/2, \pi/2)$.

Actually, the Gaussian light beam is similar to the plane wave and can be regarded as an ideal plane wave when its width of the waist is increased to be infinite. Here, a detailed analysis on the relationship between the stationary-phase approach and the Gaussian beam model is given. Supposing that the width of the incident beam is large enough, namely the half width in β space is extremely small, the reflection phase shift in the reflection coefficient $r(\beta) = |r| \exp(i\varphi)$ can be regarded as a linear function of β. On performing the Taylor expansion of φ around $\beta = \beta_0$, we obtain

$$\varphi(\beta) = \varphi(\beta_0) + \left.\frac{\mathrm{d}\varphi}{\mathrm{d}\beta}\right|_{\beta=\beta_0} (\beta - \beta_0) + o(\beta - \beta_0). \tag{7.67}$$

With the ignorance of the higher-order infinitesimal, the reflection coefficient is approximately written as

$$r(\beta) \approx |r| \exp[i\varphi'(\beta_0)] \exp\left(i \left.\frac{\mathrm{d}\varphi}{\mathrm{d}\beta}\right|_{\beta=\beta_0} \beta \right), \tag{7.68}$$

where

$$\varphi'(\beta_0) = \varphi(\beta_0) - \left.\frac{\mathrm{d}\varphi}{\mathrm{d}\beta}\right|_{\beta=\beta_0} \beta_0. \tag{7.69}$$

By substituting Eq. (7.68) into Eq. (7.66), the profile of the reflected beam is expressed as

$$\varphi_r(x, z = 0) = \frac{1}{\sqrt{2\pi}} |r| \exp[i\varphi'(\beta_0)] \times \int A(\beta) \exp\left\{ i\beta \left[x - \left(- \left.\frac{\mathrm{d}\varphi}{\mathrm{d}\beta}\right|_{\beta=\beta_0} \right) \right] \right\} \mathrm{d}\beta. \tag{7.70}$$

Compared Eq. (7.70) with Eq. (7.63), it can see that besides an added constant term $|r|$ and an added phase factor $\exp[i\varphi'(\beta_0)]$, the centric position of the reflected light distribution moves from that of the incident light distribution $x = 0$ to the position $x = -(\mathrm{d}\varphi/\mathrm{d}\beta)|_{\beta=\beta_0}$, which is identical to the GH shift defined by Eq. (7.60).

Therefore, the stationary-phase approach is the first-order approximation of the Gaussian beam model.

7.4.3 *Interference Approach [9]*

Considering the refection of a plane wave upon a prism–waveguide coupling system (see Fig. 7.9) from the viewpoint of the interference, it is shown that the reflected beam is a result of the interference between two beams: the beam directly reflected from the prism and the leaky beam coming from the guided mode. The incident wave is uniformly distributed at the base of the prism, and we only draw one light ray A_1 at x_n (shown in Fig. 7.10). In the distance between x_{n-1} and x_n (one propagation period of the guided mode), as shown in Fig. 7.10a, the light ray $(A_3)_{n-1}$ in the guiding layer is reflected into ray B_3 at the interface between medium n_3 and n_4, and ray B_3 is reflected as ray A_3' at the interface between medium n_2 and n_3 and is transmitted as ray B_1' (or leaky wave). The rays B_3' B_1', and A_3' form a self-consistent set in the vicinity of $x = x_n$, and their amplitudes are related to each other by the boundary conditions at z = d_2 and z = 0. The reflection and transmission coefficients r_{321} and t_{31} in Fig. 7.10a are obtained as

$$\begin{cases} r_{321} = \dfrac{A_3'}{B_3} = -\dfrac{r_{23} + r_{12}\exp(i2\kappa_2\mathrm{d}_2)}{1 + r_{12}r_{23}\exp(i2\kappa_2\mathrm{d}_2)}, \\ t_{31} = \dfrac{B_1'}{B_3} = \dfrac{(1 - r_{12})(1 - r_{23})(i2\kappa_2\mathrm{d}_2)}{1 + r_{12}r_{23}\exp(i2\kappa_2\mathrm{d}_2)} \end{cases}. \tag{7.71}$$

Therefore, we get $A_3' = (A_3)_{n-1}r_{34}r_{321}$ exp $(i2\kappa_3\mathrm{d}_3)$ and $B_1' = (A_3)_{n-1}r_{34}t_{31}$ exp $(i2\kappa_3\mathrm{d}_3)$. In Fig. 7.10b where the prism and the guiding layer are considered as semi-infinite medium, the incident ray A_1 is reflected as B_1'' and is transmitted as A_3'', so the rays A_1' B_1'', and A_3'' form a self-consistent set, and their amplitudes are related to each other by the boundary conditions at $z = 0$ and $z = \mathrm{d}_2$. The reflection and transmission coefficients r_{123} and t_{13} are

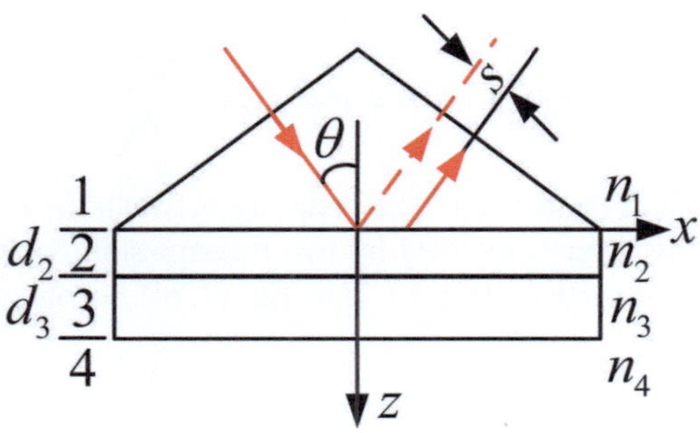

Fig. 7.9 Schematic diagram of the prism–waveguide coupling system: *1* prism, *2* coupling layer, *3* guiding layer, and *4* substrate

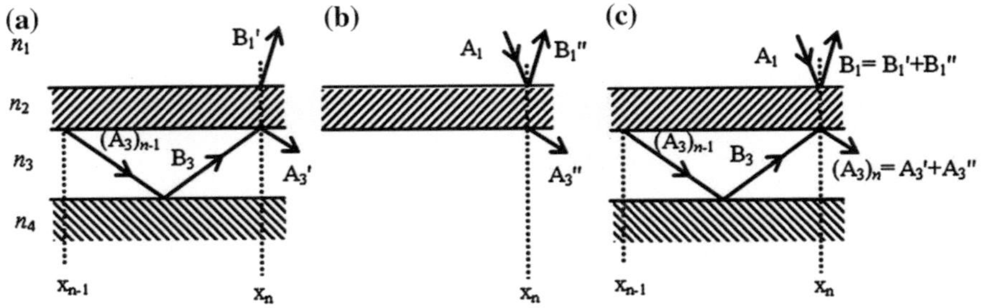

Fig. 7.10 The rays in the prism–waveguide coupling system: (**a**) the guided mode is recoupled into a semi-infinite medium n_1 (the prism) through the coupling layer n_2, (**b**) the light beam in a semi-infinite medium n_1 is coupled into another semi-infinite medium n_3 through the coupling layer n_2, and (**c**) the combination of multiple rays

$$\begin{cases} r_{123} = \dfrac{B_1''}{A_1} = \dfrac{r_{12} + r_{23}\exp(i2\kappa_2 \mathrm{d}_2)}{1 + r_{12}r_{23}\exp(i2\kappa_2 \mathrm{d}_2)}, \\ t_{13} = \dfrac{A_3''}{A_1} = \dfrac{(1 + r_{12})(1 + r_{23})(i2\kappa_2 \mathrm{d}_2)}{1 + r_{12}r_{23}\exp(i2\kappa_2 \mathrm{d}_2)}. \end{cases} \tag{7.72}$$

So $B_1'' = r_{123}A_1$ is the directly reflected ray, and $A_3'' = t_{13}A_1$ is the ray coupled into the guiding layer and becomes a part of guided mode. At the starting point of coupling x_0, which is not shown in Fig. 7.10, A_3'' is the whole guided mode $(A_3)_0$, i.e., $(A_3)_0 = t_{13}A_1$. According to Fig. 7.10c, the ray A_3 at x_n is the result of the interference between A_3' and A_3'' and is written as

$$(A_3)_n = A_3' + A_3'' = t_{13}A_1 + (A_3)_{n-1}r_{34}r_{321}\exp(i2\kappa_3 \mathrm{d}_3), \tag{7.73}$$

which is a recurrent formula with $(A_3)_0 = t_{13}A_1$. The reflected wave B_1 in Fig. 7.10c is a result of the interference between the directly reflected ray B_1'' and the total leaky wave B_1' that originates from the guided mode B_3. Therefore, the reflection coefficient of the prism–waveguide coupling system at x_n is $r = (B_1'' + B_1')/A_1$, which can be expanded as

$$r = r_{123} + r_{34}t_{31}\exp(i2\kappa_3 \mathrm{d}_3)\frac{(A_3)_{n-1}}{A_1}. \tag{7.74}$$

According to Eqs. (7.73) and (7.74), the coupling between the ray in the prism and the ray in the guiding layer is realized by two transmission coefficients t_{13} and t_{31}. By using the recurrent formula (Eq. (7.73)), Eq. (7.74) becomes

$$r = \sum_{j=0}^{n} r^j, \tag{7.75}$$

and

$$\begin{cases} r^0 = r_{123}, \\ r^j = [r_{34}r_{321}\exp(i2\kappa_3 d_3)]^{j-1} \cdot t_{13}t_{31}r_{34}\exp(i2\kappa_3 d_3), \ (j \geq 1) \end{cases} \tag{7.76}$$

where r^j is explained as follows: $r^0 A_1$ represents the directly reflected wave; part of ray A_1 in prism at x_{n-1} is coupled into guiding layer as $t_{13}A_1$, undergoing propagation from x_{n-1} to x_n in guiding layer (one propagation period) ($t_{13}A_1$ becomes $t_{13}A_1 r_{34}$ exp ($i2\kappa_3 d_3$), and part of it is recoupled into the prism as the leaky wave $r^1 A_1$; $r^j A_1$ originates from the ray A_1 in the prism at x_{n-j} and undergoes j propagation periods in the guiding layer.

By introducing Eq. (7.75) into Eq. (7.66), the field of the reflected beam is

$$\psi_r(x, z=0) = \sum_{j=0}^{\infty} \psi_r^j(x, z=0) = \sum_{j=0}^{\infty} \frac{1}{\sqrt{2\pi}} \int r^j(\beta) A(\beta) \exp(i\beta x) \mathrm{d}\beta. \tag{7.77}$$

By defining $r' = \sum_{j=1}^{\infty} r^j$, the field of the reflected beam can be rewritten as

$$\psi_r = \psi_r^0 + \psi_r', \tag{7.78}$$

where ψ_r^0 represents the field of the directly reflected beam from the base of the prism, and ψ_r' represents the field of the leaky beam originating the guided mode. It is shown that the reflected beam is the result of the interference between the multiple reflected beam constituents $\psi_r^j(x, z = 0)$.

7.5 Enhancement of the GH Shift

Usually, the GH shift is too small (on the order of wavelength) to encumber the quantification observations. Several configurations [10–14] have been proposed to enhance the GH shift. Moreover, theoretical and experimental results show that the magnitude of GH shift can be not only positive but also negative [13]. In this subsection, three of these configurations for enhancing the GH shift are described in detail.

7.5.1 *Near the Brewster Angle [10]*

According to the stationary-phase approach (see Eq. (7.60)), it indicates that a more abrupt change of the phase shift of the reflection coefficient can give rise to a larger GH shift. In the case of reflection from an interface between two transparent mediums, the real part of reflection coefficient for TM-polarized wave beam is well known

$$r = \tan(\theta - \theta_t)/\tan(\theta + \theta_t), \tag{7.79}$$

where θ_t is the angle of refraction (or transmission). Whereas the reflection coefficient vanishes at exactly the Brewster angle where $\theta + \theta_t = \pi/2$, a closer look indicates that r changes sign across the angle, and this means a sudden phase change of amount π. This infinite slope, which is accompanied by the null reflection, is of no interest. Yet in view of continuity, it is reasonable to expect a large finite slope of the change of phase with a nonzero, albeit small, reflection if the reflecting medium is weakly absorbing.The reflection coefficient for an incident TM wave reflection from an absorbing medium is written as

$$r = \frac{\varepsilon \cos\theta - \sqrt{\varepsilon - \sin^2\theta}}{\varepsilon \cos\theta + \sqrt{\varepsilon - \sin^2\theta}}, \tag{7.80}$$

where $\varepsilon = \varepsilon_r + i\varepsilon_i$ is the dielectric constant of the medium. Let us consider a weakly absorbing medium, i.e., $\varepsilon_i/\varepsilon_r \ll 1$. Therefore, r can be expanded to the first order in $\varepsilon_i/\varepsilon_r$. It is then not difficult to find that, near the Brewster angle, defined by $\theta_b = \tan^{-1}\sqrt{\varepsilon_r}$, the real part of r is given by Eq. (7.79) if $\theta_t = \sin^{-1}(\sin\theta/\sqrt{\varepsilon_r})$ is defined, whereas the imaginary part is given by

$$\mathrm{Im}(r) = \frac{\varepsilon_i \cos\theta(\varepsilon_r - 2\sin^2\theta)}{\sqrt{\varepsilon_r - \sin^2\theta}\left[\varepsilon_r \cos\theta + \sqrt{\varepsilon_r - \sin^2\theta}\right]^2}, \tag{7.81}$$

which reduces to

$$\mathrm{Im}(r_b) = \varepsilon_i(\varepsilon_r - 1)/4\varepsilon_r^2, \tag{7.82}$$

at exactly the Brewster angle. Hence for $\varepsilon_r > 1$, one has a positive imaginary part near the Brewster angle and a change of the real part from positive to negative across that angle. The phase is thus limited to the range $(0, \pi)$; it is a small positive value for $\theta < \theta_b$, increases with angle and passed through the value $\pi/2$ at $\theta = \theta_b$, and finally approaches π for $\theta > \theta_b$. The change of the phase shift of angle is definitely positive across the Brewster angle. In fact, the phase shift and the slope can be obtained from the real and the imaginary parts of the reflection coefficient in Eq. (7.79) and Eq. (7.81), respectively. At exactly θ_b, the GH shift is

$$S_b = -\frac{\varepsilon_r + 1}{\pi\sqrt{\varepsilon_r}}\frac{\varepsilon_r}{\varepsilon_i}\lambda, \tag{7.83}$$

It is negative and obviously approaches infinity in the limit of zero absorption. Furthermore, it also tells us that, for a given ratio $\varepsilon_r/\varepsilon_i$, a larger ε_r implies a grater GH shift.

7.5.2 Surface Plasmon Resonance [11]

To obtain a larger GH shift, Yin et al. [11] described a surface plasmon resonance configuration, with which a GH shift of greater than 50 wavelengths was observed, since much more light energy was coupled into the medium under the metal film. The Kretschmann's configuration for the surface plasma excitation is shown in Fig. 7.11. The dielectric constants of the air, thin metal film, and the prism are denoted as ε_0, ε_1, and ε_2, respectively, and d is the thickness of metal thin film. As a TM-polarized light beam incidents upon the interface between prism and metal and the resonance condition is satisfied, the surface plasmon wave will be excited. In the experiment, the position of reflected beam is detected by a position-sensitive detector (PSD). By periodically modulating the incident polarization, the difference of the lateral displacement between TE light and TM light is measured. However, since the TE-polarized incidence cannot excite any surface plasmon resonance, there is no enhanced GH shift. Therefore, it serves as a perfect reference beam, and the measured relative beam shift between TM and TE excitation indeed indicates an absolute beam displacement (see the solid line in Fig. 7.11) for a TM wave at the surface plasmon resonance region.

As scanning the incident angle θ, the minimal reflection intensity $R_{\min}$ in the attenuated-total-reflection curve is closely related to the thickness of the thin metal film. Experimental results show the following:

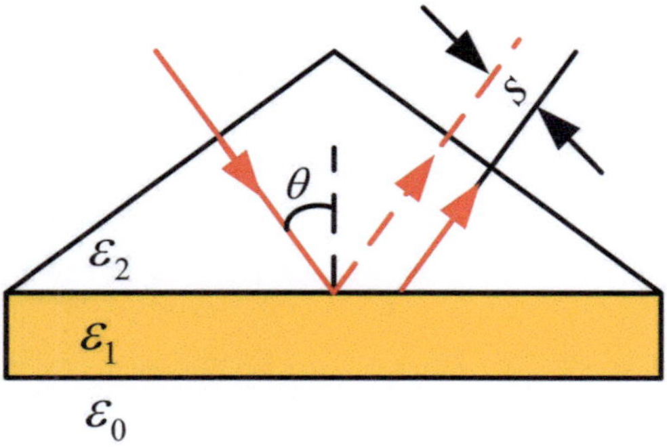

Fig. 7.11 Kretchmann's configuration for the surface plasma excitation

(i) There is a critical thickness h_c for the thin metal film. As the thickness of thin metal film approaches to that critical thickness, $R_{\min}$ decreases, but the GH shift increases significantly. A GH shift of as high as 50 wavelengths is observed;
(ii) The GH shift is positive if the thickness of thin metal film is smaller than the critical thickness $h < h_c$. Additionally, the GH shift becomes negative for thin metal films thicker than the critical thickness $h > h_c$.

7.5.3 Prism–Waveguide Coupling System [12]

Let us consider the prism–waveguide coupling system shown in Fig. 7.12. The dielectric constants of the substrate, the guiding layer, the air gap, and the prism are denoted as ε_0, ε_1, ε_2, and ε_3, respectively. The thickness of the guiding layer and the air gap is d_1 and d_2, respectively. Through the analysis described in Chap. 4, when the phase-matching condition is satisfied, we get Eq. (4.39)

$$\beta = \mathrm{Re}(\beta^0) + \mathrm{Re}(\Delta\beta^L),$$

and the minimum of reflectivity (Eq. (4.44))

$$R_{\min} = |r_{32}|^2 \left\{ 1 - \frac{4\mathrm{Im}(\beta^0)\mathrm{Im}(\Delta\beta^L)}{\left[\mathrm{Im}(\beta^0) + \mathrm{Im}(\Delta\beta^L)\right]^2} \right\}.$$

According to the stationary-phase approach [2], the GH shift of the prism–waveguide coupling system is approximately expressed as [12]

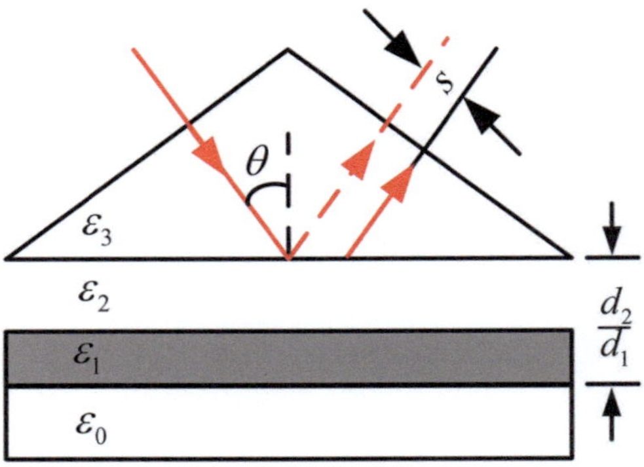

Fig. 7.12 The prism–waveguide coupling system

$$S = -\frac{2\mathrm{Im}(\Delta\beta^L)}{\mathrm{Im}(\beta^0)^2 - \mathrm{Im}(\Delta\beta^L)^2}\cos\theta. \tag{7.84}$$

By applying Eq. (7.84), we obtain two important conclusions:

(i) When the intrinsic damping is well matched with the radiative damping, namely

$$\mathrm{Im}(\beta^0) = \mathrm{Im}(\Delta\beta^L), \tag{7.85}$$

the corresponding GH shift will be maximal.

(ii) The sign of the GH shift is determined by the intrinsic and radiative dampings. When the intrinsic damping is larger than the radiative damping $\mathrm{Im}(\beta^0) > \mathrm{Im}(\Delta\beta^L)$, a negative GH shift $S < 0$ can be obtained. The positive GH shift $S > 0$ corresponds to the reverse case, $\mathrm{Im}(\beta^0) < \mathrm{Im}(\Delta\beta^L)$. Since the radiative damping mainly depends on the thickness of the air gap, there is also a critical thickness in the prism–waveguide coupling system. The sign of GH shift dependence on the comparison between the thickness of air gap and the critical thickness [13] is identical to that of the surface plasmon resonance.

7.5.4 Symmetrical Metal-Cladding Waveguide [14]

The schematic diagram of the symmetrical metal-cladding waveguide (SMCW) is shown in Fig. 6.1, which includes three layers: an upper metal film serving as the upper cladding as well as the coupling layer, a glass slab with millimeter thickness acting as the guiding layer, and a relatively thick metal film deposited upon the other side of the glass slab working as the substrate. When Eq. (7.85) is satisfied in the SMCW, the GH shift of the reflected light beam will be greatly enhanced.

The SMCW structure used in the experiment is a glass slab with two silver films coated on both the upper and lower sides. Their thicknesses are 3.04 mm, 33.1, and 300 nm, respectively. The schematic diagram of the experimental arrangement is shown in Fig. 7.13. A TE-polarized Gaussian beam emitted from a tunable laser was introduced onto the SMCW. Before it reaches the coupling layer, the beam passes through several devices: two apertures with a diameter of 2 mm and a distance of 0.5 m, which are used to further confine the divergence of the incident light; a beam splitter which is used to reject part of the beam into a wavemeter for wavelength monitoring. The intensity of the reflected beam is detected by a photodiode (PD). The angular scanning is carried out by a computer-controlled $\theta/2\theta$ goniometer. The incident angle is selected to be located at the maximum reflectivity of one certain reflectivity dip, where the GH shift is not remarkable. Since the magnitude of the GH shift is strongly dependent on the energy coupling between the incident light and the guided mode, it is reasonable to take this position of the

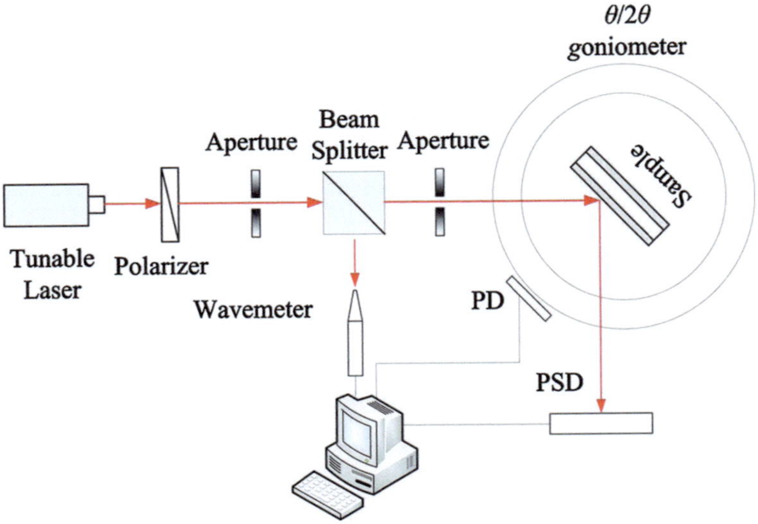

Fig. 7.13 Experimental setup for enhancing GH shift

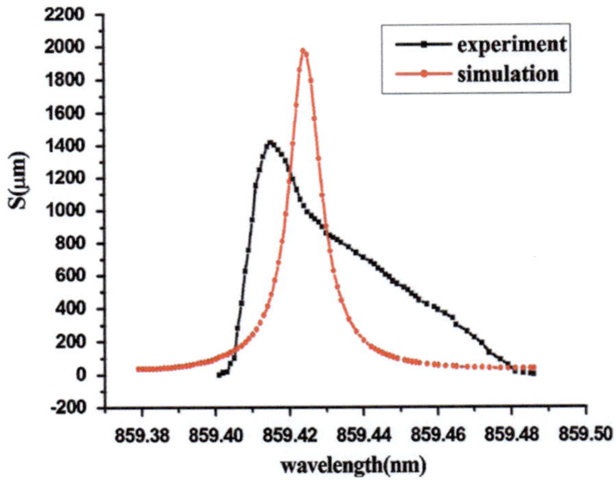

Fig. 7.14 Theoretical and experimental reflected beam displacements (S) with respect to the light wavelength

reflected beam as the reference of the GH shift. After removing the PD out of its position without changing any position of the incident beam and the structure, we put on the position-sensitive detector (PSD) and let the reflected beam impinge onto the PSD at the center perpendicularly. Then, we change the wavelength of the incident light from 859.401 to 859.415 nm and get a tremendous GH shift greater than 1700 wavelengths, which to our knowledge is the largest result in experiment.

The GH shift of the reflected beam changes with the variation in the wavelength as shown in Fig. 7.14. The simulation in the Gaussian beam model is also outlined for comparison. The differences between the experimental observation and theoretical simulation most probably result from slight non-parallelism of the guiding layers, instability of the laser, the defection of the silver films, and errors in the measurements of the structure parameters. The result shows that the GH shift as large as 1.5 mm is obtained with the working wavelength changing only by 14 pm.

7.6 Other Non-Specular Reflection Effects

It is well known that the interaction of a plane wave with an interface is described by the Snell's law and the Fresnel formulae. For a real optical beam which has a finite width (i.e., a distributed plane wave spectrum), there are four non-specular effects deviating from the geometrical optics picture. As shown in Fig. 7.15, besides the GH shift, other non-specular shifts are the Imbert–Fedorov (IF) shift, (i.e., a spatial shift perpendicular to the plane of incidence), the angular GH shift, and the angular IF shift, respectively.

Consider a system consisting of two homogeneous isotropic mediums of dielectric constants ε_1 and ε_2 filling the half spaces $z < 0$ and $z \geq 0$, respectively. A monochromatic beam of light of wavelength λ and waist w_0 propagates along the central wave vector $\vec{k}_0$ in the region $z < 0$ before impinging upon the plane interface of equation $z = 0$ that separates medium 1 from medium 2. The expressions for four non-specular effects can be written as [15, 16]

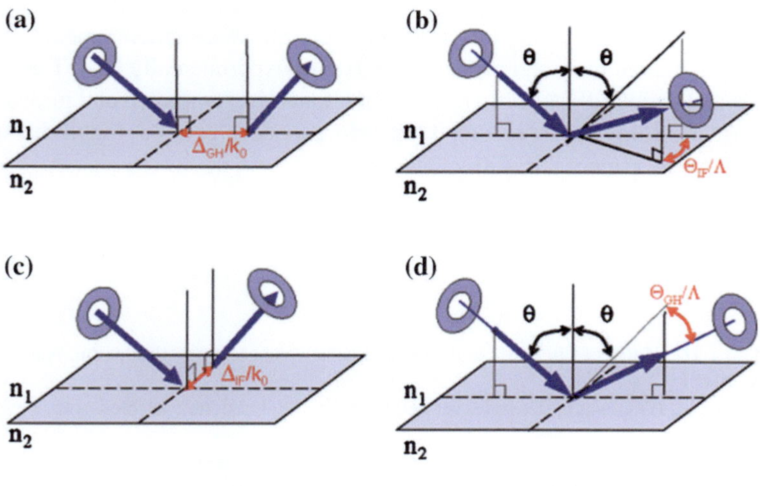

Fig. 7.15 Cartoonlike representation of the four non-specular effects: **a** the Goos–Hänchen shift, **b** the angular Imbert–Fedorov shift, **c** the Imbert–Fedorov shift, and **d** the angular Goos–Hänchen shift

$$\begin{cases} \Delta_l = \lambda_0 \mathrm{Im}[D_l], \\ \Theta_l = -(\theta_0^2/2)\mathrm{Re}[D_l], \end{cases} \tag{7.86}$$

where Δ and Θ are the spatial and angular shifts, respectively, the index l is a label for the two linear polarizations parallel ($l = p$ or TM) and perpendicular ($l = s$ or TE) to the central plane of incidence $x - z$, $\lambda_0 = \lambda/2\pi$, and $\theta_0 = 2\lambda_0/w_0$ is the angular spread of the incident beam. Equation (7.86) is valid for both the GH and the IF shifts where the coefficient D_l is equal to

$$D_l = \frac{\partial \ln r_l}{\partial \theta} = \frac{1}{R_l}\frac{\partial R_l}{\partial \theta} + i\frac{\partial \phi_l}{\partial \theta}, \tag{7.87}$$

in the GH case and is equal to

$$D_l = 2i\left[(r_p + r_s)/r_l\right]\cot\theta, \tag{7.88}$$

in the IF case. In the latter case, the results calculated from Eq. (7.86) are the spatial and angular separation between the two right circularly and left circularly polarized components of the reflected beam generated by the reflection-induced splitting of the l–polarized $l = \{p, s\}$ incident beam. Moreover, $r_l \equiv r_l(\theta) = R_l \exp(i\phi_l)$ is the Fresnel reflection coefficient evaluated at the central incident angle θ, where $R_l = |r_l|$ and $\phi_l = \arg r_l$.

Note that the studies of IF shift reveal more deep and sophisticated physics. The IF shift has been connected with the angular momentum conservation [17] and the spin Hall effect of light [18, 19]. Usually, the magnitude of the IF shift is one order smaller than that of the GH shift. In 2008, the weak measurement approach was presented to amplify and measure the IF shift experimentally [20]. The angular GH shift and angular IF shift can also be simultaneously amplified and measured by the weak measurement approach [21]. More details can be found in several excellent review papers [22, 23].

References

1. F. Goos, H. Hänchen, Ein neuer und fundamentaler versuch zur totalreflexion. Ann. Phys. **1**, 333 (1947)
2. K. Artmann, Berechnung der seitenversetzung des totalreflektierten strahles. Ann. Phys. **2**, 87 (1948)
3. K.J. Resch, J.S. Lundeen, A.M. Steinberg, Total reflection cannot occur with a negative delay time. J. Quantum Electron. **37**, 794 (2001)
4. P. Tournois, Negative group delay times in frustrated Gires-Tournois and Fabry-Perot interferometers. J. Quantum Electron. **33**, 519 (1997)
5. P. Tourmois, Apparent causality paradox in frustrated Gires-Tournois interferometers. Opt. Lett. **30**, 815 (2005)

6. X.M. Liu, Z.Q. Czo, P.F. Zhu, Q.S. Shen, Time delay associated with total reflection of a plane wave upon plasma mirror. Opt. Express **14**, 3588 (2006)
7. X.M. Liu, Z.Q. Cao, P.F. Zhu, Q.S. Shen, Solution to causality paradox upon total reflection in optical planar waveguide. Phys. Rev. E **73**, 016615 (2006)
8. C.F. Li, Q. Wang, Prediction of simultaneously large and opposite generalized Goos-Hänchen shifts for TE and TM light beams in an asymmetric double-prism configuration. Phys. Rev. E **69**, 055601 (2004)
9. X.M. Liu, Q.F. Yang, Z. Qiao, T.K. Li, P.F. Zhu, Z.Q. Cao, Physical origin of large positive and negative lateral optical beam shifts in prism-waveguide coupling system. Opt. Commun. **283**, 2681 (2010)
10. H.M. Lai, S.W. Chan, Large and negative Goos-Hänchen shift near the Brewster dip on reflection from weakly absorbing medium. Opt. Lett. **27**, 680 (2002)
11. X.B. Yin, L. Hesselink, Z.W. Liu, N. Fang, X. Zhang, Large positive and negative lateral optical beam displacement due to surface plasmon resonance. Appl. Phys. Lett. **85**, 372 (2004)
12. X.B. Liu, Z.Q. Cao, P.F. Zhu, Q.S. Shen, X.M. Liu, Large positive and negative lateral optical beam shift in prism-waveguide coupling system. Phys. Rev. E **73**, 056617 (2006)
13. L. Chen, Z. Cao, F. Ou, H. Li, Q. Shen, H. Qiao, Observation of large positive and negative lateral shifts of a reflected beam from symmetrical metal-cladding waveguides. Opt. Lett. **32**, 1432 (2007)
14. J. Hao, H. Li, C. Yin, Z. Cao, 1.5 mm light beam shift arising from 14 pm variation of wavelength. J. Opt. Soc. Am. B **27**, 1305 (2010)
15. W. Nasalski, T. Tamir, L. Lin, Displacement of the intensity peak in narrow beams reflected at a dielectric interface. J. Opt. Soc. Am. A **5**, 132 (1988)
16. A. Aiello, M. Merano, J.P. Woerdman, Duality between spatial and angular shift in optical reflection. Phys. Rev. A **80**, 061801 (2009)
17. M.A. Player, Angular momentum balance and transverse shifts on reflection of light. J. Phys. A Math. Gen. **20**, 3667 (1987)
18. M. Onoda, S. Murakami, N. Nagaosa, Hall effect of light. Phys. Rev. Lett. **93**, 083901 (2004)
19. K. Yu Bliokh, Y.P. Bliokh, Conservation of angular momentum, transverse shift, and spin Hall effect in reflection and refraction of an electromagnetic wave packet. Phys. Rev. Lett. **96**, 073903 (2006)
20. O. Hosten, P. Kwiat, Observation of the spin hall effect of light via weak measurement. Science **319**, 787 (2008)
21. S. Goswami, M. Pal, A. Nandi, P.K. Panigrahi, N. Ghosh, Simultaneous weak value amplification of angular Goos-Hänchen and Imbert-Fedorov shifts in partial reflection. Opt. Lett. **39**, 6229 (2014)
22. C. Leyder, M. Romanelli, JPh Karr, E. Giacobino, T.C.H. Liew, M.M. Glazov, A.V. Kavokin, G. Malpuech, A. Bramati, Observation of the optical spin hall effect. Nat. Phys. **3**, 628 (2007)
23. K.Y. Bliokh, Geometrodynamics of polarized light: barry phase and spin Hall effect in a gradient-index medium. J. Opt. A **11**, 094009 (2009)

Chapter 8
Optical Devices Based on the Attenuated Total Reflection

Abstract This chapter introduces several basic optical devices based on the attenuated total reflection, including tunable filter, optical sensors, and electro-optical devices. We mainly focus on the description of optical sensors performance. It is found that the bigger the portion of power that propagates in the sample, the higher the sensitivity will be. Moreover, the sensors based on the GH shift are immune to the power fluctuation in the light source since the GH shift is position encoded. Finally, our experiments to explore the magneto-optical modulation and all-optical modulation in the ferrofluid-filled SMCW are given. The results are contributed to the competition between the optical trapping effect and the Soret effect.

Keywords Attenuated total reflection · Filter · Optical sensor · Oscillating wave · Ferrofluid

8.1 Optical Waveguide Filters

Optical filters play an important role in the fields of optoelectronics and optical telecommunications. However, most of the available optical filters are polarization dependent. Furthermore, the dense wavelength-division multiplexing (DWDM) has attracted considerable interest since it can send many independent channels in one optical fiber simultaneously. In such a DWDM system, a comb filter is a key component to decrease the cross talk between the neighboring channels. So far, various technologies for generating a comb filter have been reported but rare can work with the channel spacing down to 0.8 or 0.4 nm. In this section, we firstly describe a tunable narrow band filter possessing a capability of the polarization-insensitive operation and then propose a tunable comb filter based on a symmetrical metal-cladding waveguide (SMCW) structure working with the channel spacing of 0.8 nm.

X. Wang et al., *Progress in Planar Optical Waveguides*,
Springer Tracts in Modern Physics 266, DOI 10.1007/978-3-662-48984-0_8

8.1.1 *Tunable Narrow Band Filter [1]*

The structure of the tunable narrow band filter, which is built by a glass slab with two gold films deposited on its upper and bottom sides, is illustrated in Fig. 8.1. The thickness of the glass slab is of a submillimeter scale. The upper gold film of 30 nm acts as a coupling layer, and the bottom one of 200 nm functions as a substrate. The dielectric constant of the air, the glass slab, and the gold film is denoted as ε_0, ε_1, and ε_2, respectively.

If the filter is illuminated by a monochromatic light, a series of resonant dips will be formed in the reflectivity spectrum because the energy of light with certain incident angles can be free-space coupled into the glass slab. These resonant dips have been proven to be polarization insensitive [2]. On the other hand, if a polychromatic light is used and the incident angle is fixed, the resonant dips in the reflectivity spectrum can also be excited by scanning the incident wavelength.

The experimental arrangement for obtaining a tunable narrow spectral resonance bandwidth and a polarization-insensitive filter is shown in Fig. 8.2. A tunable diode laser, whose precise wavelength tuning is implemented by a current control unit, is operated at the central wavelength of 859 nm with a laser linewidth of 1 MHz. After passing through a polarizer and a quarter-wave plate, the output light beam becomes circularly polarized. A splitter is used to reflect a small portion of light to a wavelength measurement system. A great portion of light is incident on the upper gold film of the SMCW with an appropriate angle. Reflected light is divided into equal TE- and TM-polarized components by a beam polarized splitter and recorded by two detectors independently.

The calculated and experimental results of the normalized reflected intensity as a function of wavelength are given in Fig. 8.3. The experimentally measured spectral resonance bandwidth of the full width at half maximum (FWHM) is 0.08 nm. The deviation between the calculation and the experimental results is mainly originated from the beam divergence of the laser beam. Note that the reflection spectrum is identical for both the TE and the TM modes; namely, the filter is polarization insensitive.

The tunability of the filter can easily be obtained by slight rotation of the sample with respect to the incident light beam. Different from the prism–waveguide coupling system, the free-space coupling technology offers a maximum angular range

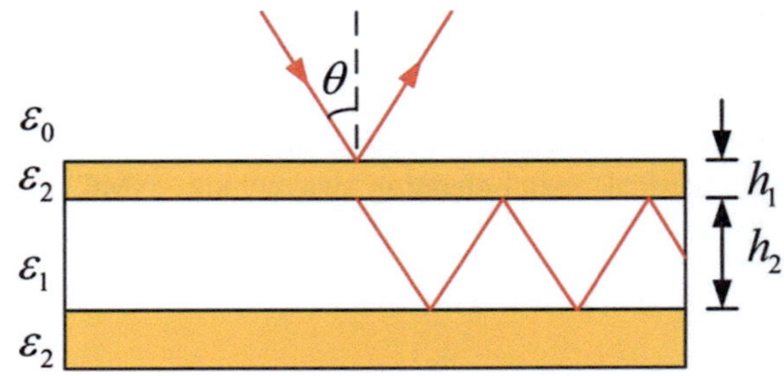

Fig. 8.1 Structure of the tunable narrow band filter

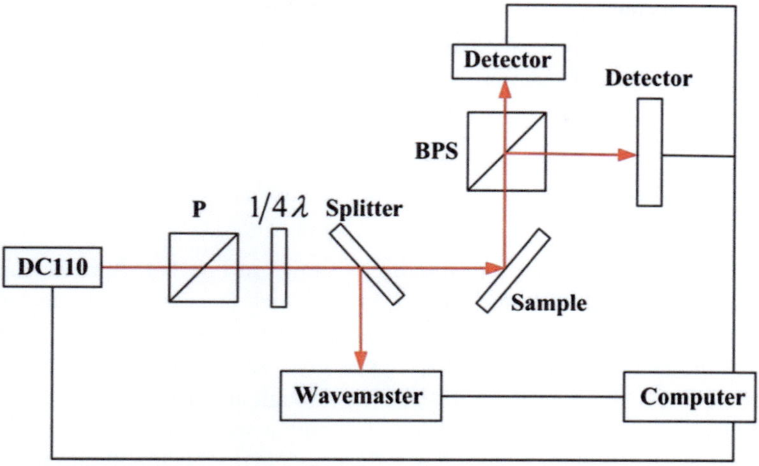

Fig. 8.2 Experimental arrangement of the tunable narrow band filter

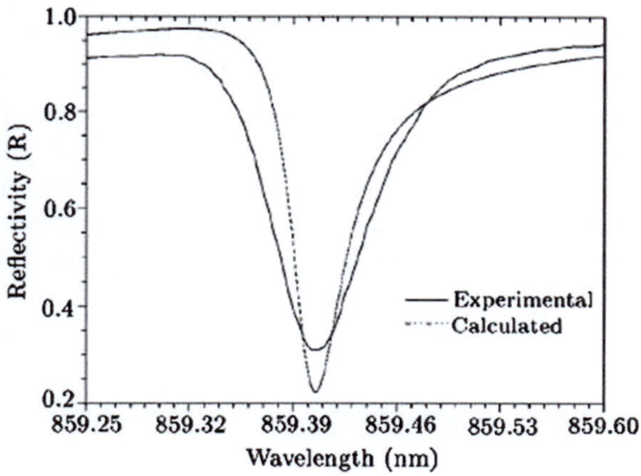

Fig. 8.3 Normalized reflected intensity as a function of wavelength

which can be varied from the normal to graze direction with respect to the waveguide plane. According to the dispersion Eq. (6.1) of the SMCW and $n_{\text{eff}} = n_0 \sin \theta_0$, one can obtain an approximate relationship between two peak wavelengths of the filter at the normal reflection ($\theta_0 = 0°, \lambda_0$) and at the graze reflection ($\theta_0 = 90°, \lambda$)

$$\frac{\lambda_0}{\lambda} = \sqrt{\frac{\varepsilon_1}{\varepsilon_1 - \varepsilon_0}}. \tag{8.1}$$

The ratio of the maximal peak length to the minimal peak length is 1.34 in case of $\varepsilon_1 = 2.25$ and $\varepsilon_0 = 1.0$. It indicates that the tuning range can be larger than 100 nm. In addition, Eq. (6.33) reveals that the bandwidth ($\Delta\lambda$) of the filter is proportion to the effective refractive index (*N*). Based on the strong dispersion effect of the ultrahigh-order mode, the bandwidth of the filter can be further narrowed by exciting more large order mode.

8.1.2 Tunable Comb Filter [2]

The experimental setup for comb filtering is shown in Fig. 8.4. A glass slab with thickness 900 μm is used as the guiding layer, which is sandwiched between two gold films. When the light generated from a wideband source passes through a self-focusing lens and irradiates the upper gold surface of the SMCW at an appropriate angle, the reflected light, passing through another self-focusing lens, is detected by an optical spectrum analyzer. The output spectrum from the comb filter is depicted by the solid curve (see Fig. 8.5), which agrees well with the simulation result (dashed curve).

The dispersion equation of the SMCW can be simplified as

$$\frac{2\pi v}{c} n_1 h_1 \cos\theta_1 = m\pi, \tag{8.2}$$

where n_1 and h_1 are the RI and the thickness of the guiding slab, respectively. θ_1 is the incident angle of the light rays inside the waveguide. v and c are the frequency and the velocity of the light in free space, respectively. m is the mode order. Without considering the material dispersion of the waveguide, we can express the frequency spacing between two neighboring channels as

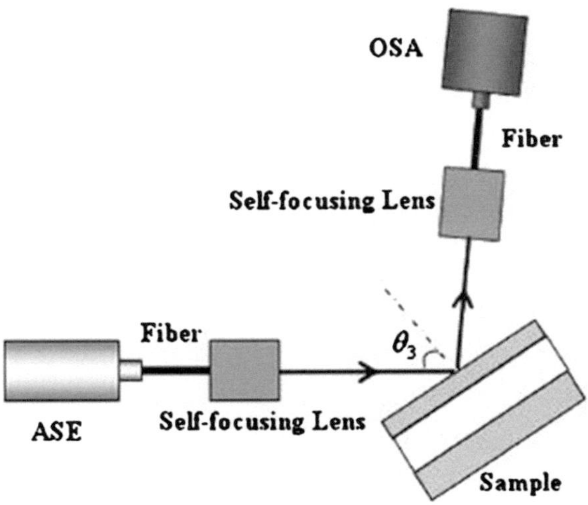

Fig. 8.4 Experiment setup of comb filter

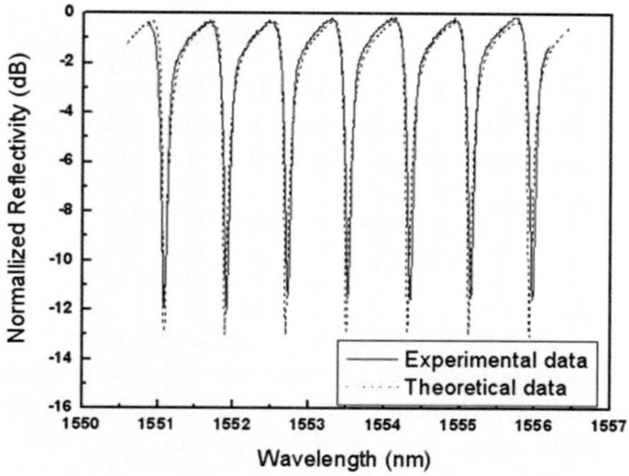

Fig. 8.5 Experimental and theoretical reflective spectrum with parameters: $\varepsilon_1 = 3.115$, $\varepsilon_2 = -132 + i12.56$, h_2 = 21 nm, h_1 = 900 μm, and θ_0 = 6.1°

$$\Delta v = \frac{c}{2n_1 h_1 \cos\theta_1}, \tag{8.3}$$

which is a constant when the incident angle of the light is fixed. In other words, all the channel frequency spacings are equal in this comb filter. Furthermore, based on Eqs. (8.2) and (8.3), we find that both the center wavelength and the channel spacing can be tuned by simply changing the incident angle. The experimental results show that the channel isolation is larger than 12 dB, the insertion loss is lower than 0.2 dB, and the wavelength spacing of loss peaks is just 0.8 nm.

8.2 Analysis on the Sensitivity

The sensitivity is defined as the magnitude of sensor transduction signal change in response to the change in analyte. Physically, the sensitivity is determined by the strength of light–matter interaction. Therefore, it is an important parameter to evaluate the sensor performance.

8.2.1 Definition of Sensitivity

Sensitivity is the ratio of the change in sensor output to the change in the quantity to be measured. Its expressions are closely related to the modulation methods. Here, two of them are listed as follows:

1. Angle modulation method
   The variation in the quantity to be measured will lead to a collective movement of the resonance dip and the corresponding resonance angle. Therefore, the resonance angle can be used as a signal to determine the quantity to be measured. In such angle modulation method, the sensitivity is written as

$$s = \frac{\partial}{\partial y}\left(\frac{\partial R}{\partial \theta}\right) = \left(\frac{\partial^2 R}{\partial \theta^2}\right)_{\theta=\theta_r} \cdot \left(\frac{\partial \theta}{\partial y}\right), \tag{8.4}$$

   where θ_r is the position of the resonant angle, and y denotes the quantity to measured (e.g., the RI, the absorption, and the thickness). The first term describes the second derivative of the reflectivity with respect to the incident angle and depends on the shape of the reflectivity curve. The second term is connected with the sensing efficiency, whose physical meaning will be fully detailed in the next section. It is clear that if the reflectivity curve has a sharper slope or a smaller FWHM, its sensitivity will be higher.
2. Intensity modulation method
   In this method, the incident angle is fixed at the middle point of the falling or the rising edge of the resonant dip, where a good linearity and a high sensitivity can be achieved. A change in the quantity to be measured will lead to a movement of the reflectivity curve, which gives rise to a change in the reflected intensity. The sensitivity formula is expressed as

$$s = \frac{\partial R}{\partial y} = \left(\frac{\partial R}{\partial \theta}\right)_{\theta=\theta_s} \cdot \left(\frac{\partial \theta}{\partial y}\right), \tag{8.5}$$

   where θ_s is the fixed incident angle, and y is the quantity to be measured. The sensitivity also consists of two terms: The first one can be understood as the sharpness of the falling or the rising edge of the resonant dip, and the second one is also connected with the sensing efficiency. From Eq. (8.5), it is obvious that the sensor's sensitivity can be improved by minimizing the width of the resonance dip.

8.2.2 Physical Meaning of the Sensing Efficiency

Herein, we discuss the physical meaning of $\partial\theta/\partial y$ in the above two definitional expressions of sensitivity. The parameter $\partial\theta/\partial y$ denotes the resonance angle shift of the whole resonance dip caused by the change in the quantity to be measured. Obviously, it is equivalent to the variation in the effective RI of the resonant mode responding to the changes in the quantity to be measured. Considering the three-layer planar waveguide shown in Fig. 8.6 and taking the TE mode as an example, we obtain

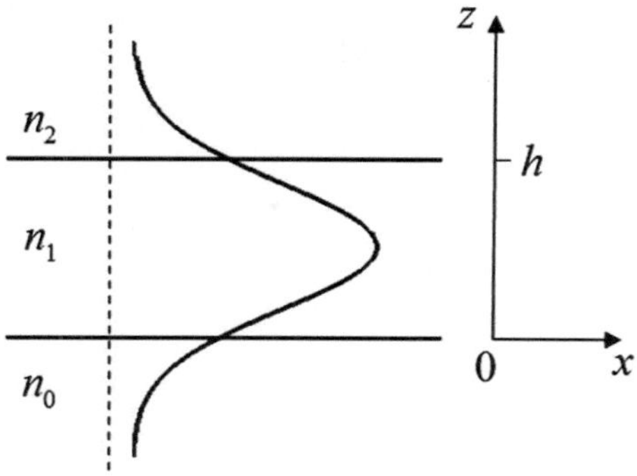

Fig. 8.6 The three-layer planar waveguide and its field distribution of TE_0 mode

$$\frac{\mathrm{d}N}{\mathrm{d}n_1}=\frac{n_1}{N}\cdot\frac{h+\frac{p_0}{\kappa_1^2+p_0^2}+\frac{p_2}{\kappa_1^2+p_2^2}}{h+\frac{1}{p_0}+\frac{1}{p_2}}=\frac{n_1}{N}\cdot\frac{P_1}{P}, \tag{8.6}$$

$$\frac{\mathrm{d}N}{\mathrm{d}n_2}=\frac{n_2}{N}\cdot\frac{\kappa_1^2}{p_2\left(\kappa_1^2+p_2^2\right)\left(h+\frac{1}{p_0}+\frac{1}{p_2}\right)}=\frac{n_2}{N}\cdot\frac{P_2}{P}, \tag{8.7}$$

$$\frac{\mathrm{d}N}{\mathrm{d}n_0}=\frac{n_0}{N}\cdot\frac{\kappa_1^2}{p_0\left(\kappa_1^2+p_0^2\right)\left(h+\frac{1}{p_0}+\frac{1}{p_2}\right)}=\frac{n_0}{N}\cdot\frac{P_0}{P}, \tag{8.8}$$

where N is the effective RI, P, P_1, P_2, and P_0 represent the power flowing in the whole waveguide structure, in the guiding layer, in the cladding layer, and in the substrate, respectively. The sensing efficiency is proportional to the ratio of the power in the sensing area to the power in the whole waveguide structure. Furthermore, it is closely connected with the ratio of the RI of the measured medium to the effective RI. For the evanescent wave sensors, N is always larger than n_0 and n_2, i.e., $n_0/N<1$ and $n_2/N<1$. But for the SMCW-based oscillating wave sensor, the allowed range of the effective RI is $0<N<n_1$, which makes it possible to obtain an enormously high sensitivity on condition that the ultrahigh-order modes ($N\rightarrow 0$) at small incident angles are used as the sensing probe.

8.3 Evanescent Wave Sensors

Optical sensors are a powerful detection and analysis tool in several fields, including biomedical research, environmental monitoring, and pharmaceuticals. Generally, there are two categories of optical wave, i.e., the evanescent wave and the oscillating wave, which can be employed to interact with the analyte. The surface plasmon resonance (SPR) sensor [3–5], the leaky waveguide sensor, and the reverse symmetry waveguide fall into the first category, and the symmetrical metal-cladding waveguide (SMCW) sensor belongs to the second one. All above-mentioned optical sensors are immune to the electromagnetic interference.

8.3.1 SPR Sensor

As described in Chap. 5, SPR is a charge density oscillation that may exist at the interface of two medium with dielectric constants of opposite signs, for instance, a metal and a dielectric. In 1968, Otto and Kretschmann independently observed the excitation of surface plasmon as a drop in the reflectivity in two different configurations by using the attenuated-total-reflection (ATR) method. Owing to the fact that the vast majority of the field of a surface plasmon is concentrated in the dielectric medium, the propagation constant is extremely sensitive to changes in the dielectric medium. This property is the underlying physical principle of the SPR sensors. The Kretschmann's configuration for exiting surface plasmon is shown in Fig. 8.7. A light wave is totally reflected at the interface between a prism coupler and a thin metal layer, and the surface plasmon is excited at the outer boundary of the metal by evanescently tunneling through the thin metal layer.

A tiny change in real part or imaginary part of the dielectric constant of the sample will lead to a measurable variation in the resonance dip of the ATR spectrum. As depicted in Fig. 8.8, a resonant angle shift can be attributed to the change in real part and a variation in the minimum reflectance is resulted from the change in imaginary part, respectively. By measuring these variations in the resonance dip, many biochemical processes can be monitored, including the antibody–antigen interactions, the preparation of Langmuir–Blodgett membrane, and so on. However, since the resonant dip possesses a flat bottom and a large FWHM, typically about 0.5–1 degree, it is very difficult to distinguish the resulted minute shift in resonant

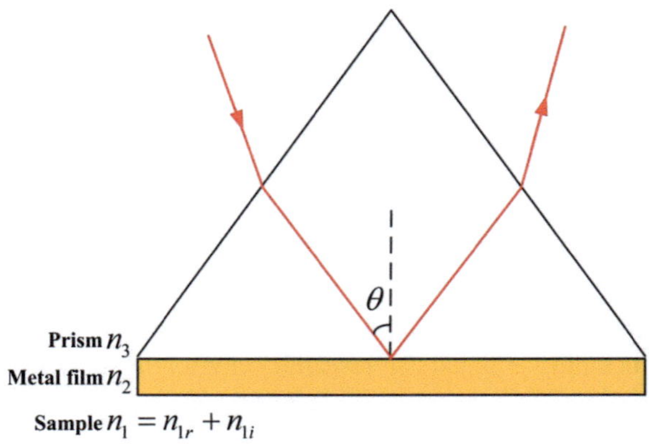

Fig. 8.7 Kretschmann's configuration for exiting surface plasmon

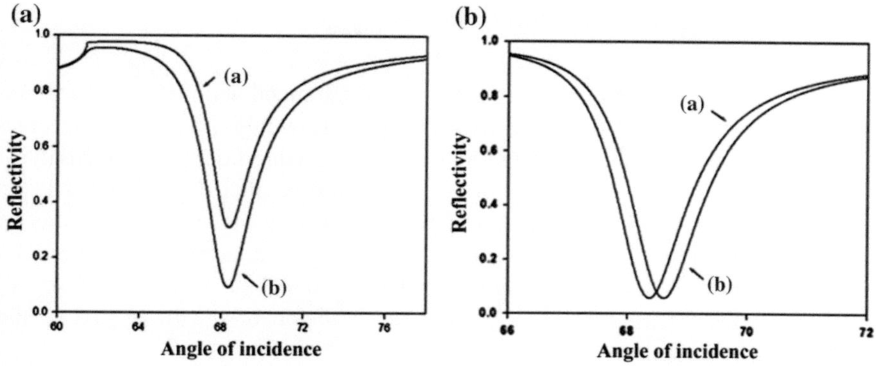

Fig. 8.8 The change of reflectivity with respect to **a** the variation in the imaginary part of dielectric constant: *sample A* $n_{1i} = 0$ and *sample B* $n_{1i} = 0.002$ and **b** the variation in the real part of dielectric constant: *sample A* $n_{1r} = 1.330$ and *sample B* $n_{1r} = 1.332$, respectively. Other simulated parameters are as follows: a BK7 glass prism $n_3 = 1.515$, a silver film $n_2 = 0.0666 + i4.045$, and wavelength $\lambda = 632.8\,\text{nm}$

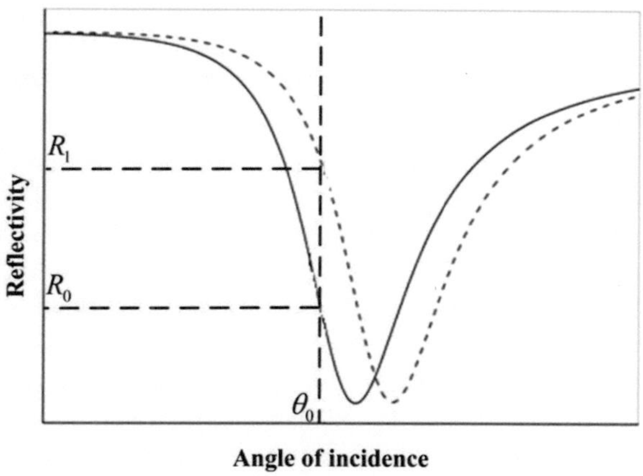

Fig. 8.9 The reflectivity variation responding to the shift of resonant dip

angle. As shown in Fig. 8.9, the intensity modulation method is an appropriate method to enhance the sensitivity. The incident angle is fixed at the middle point of the falling or the rising edge (θ_0), and when the real part of the dielectric constant changes, the resonant dip moves to the right or left side and the reflectivity changes from R_0 to R_1. Once the reflectivity variation $\Delta R = R_1 - R_0$ has been measured, the variation in real part of the dielectric constant can be determined. The slope of the falling or the rising edge is larger than that of the bottom, and therefore, the sensitivity in this method can be greatly enhanced.

8.3.2 The Leaky Waveguide Sensor

The SPR propagates along the interface between the metal and dielectric medium, but the noble metals can absorb the visible light considerably. As a result, the small slope ($dR/d\theta$) of ATR along the falling or the rising edge limits the sensitivity of SPR sensors. To increase the sensitivity, Okamtoto et al. [6] proposed a leaky waveguide sensor, whose structural configure is shown in Fig. 8.10. It contains no metal film and composes of a glass prism, a cladding layer, a guiding layer, and a sample substrate. As the light beam illuminates around the vicinity of resonant angle, the reflectivity curve of ATR is a sensitive function with respect to the complex RI of sample.

Besides the capability of detecting the concentration of the liquid sample, the leaky waveguide sensor can be used to measure the absorption, namely the extinction coefficient $\mathrm{Im}(n_3)$ of the liquid solution. The simulated reflectivity as a function of extinction coefficient is given in Fig. 8.11. The corresponding experiment has been done by Okamoto, who demonstrated that the sensitivity of the leaky waveguide sensor is larger than that of SPR sensors.

8.3.3 Reverse Symmetry Waveguide Sensor

As demonstrated in Sect. 8.2.2, a bigger proportion of light energy interacting with the sample would bring a higher sensitivity. In the conventional waveguide geometry, the substrate has a RI higher than that of the cladding layer, which usually consists of water and has a RI of 1.33. Therefore, the waveguide mode has a light intensity more concentrated toward the substrate layer, thus leaving less light to interact with the analyte. As shown in Fig. 8.12b, in a reverse symmetry waveguide [7, 8], nanoporous silica is used as the substrate. Nanoporous silica has a RI of 1.193, much lower than that of water. As a result, more light can be concentrated near the sensing surface to enhance the sensor sensitivity. Furthermore, the penetration depth of evanescent wave in the reverse symmetry waveguide sensor is bigger than that of the conventional waveguide, and consequently, the

Fig. 8.10 The structure of leaky waveguide sensor

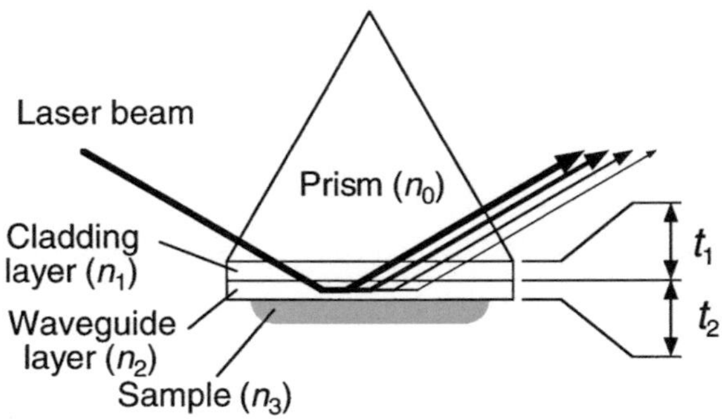

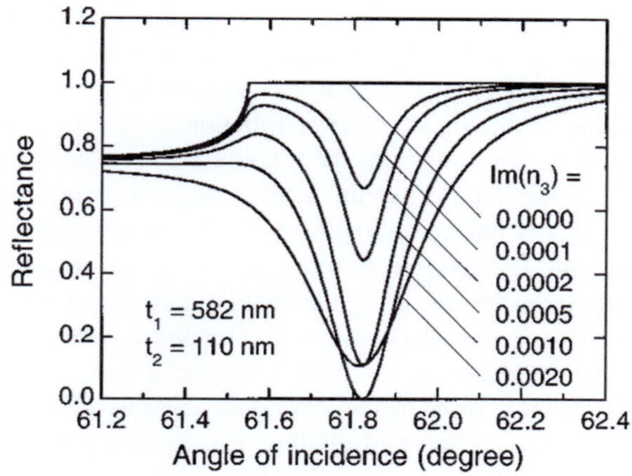

Fig. 8.11 The reflectivity curves with different extinction coefficients as a function of incident angle and other simulated parameters are $n_0 = 1.515$, $n_1 = 1.303$, $n_2 = 1.490$, $n_3' = 1.332$, $d_1 = 110\,\text{nm}$, $d_2 = 582\,\text{nm}$, and $\lambda = 632.8\,\text{nm}$

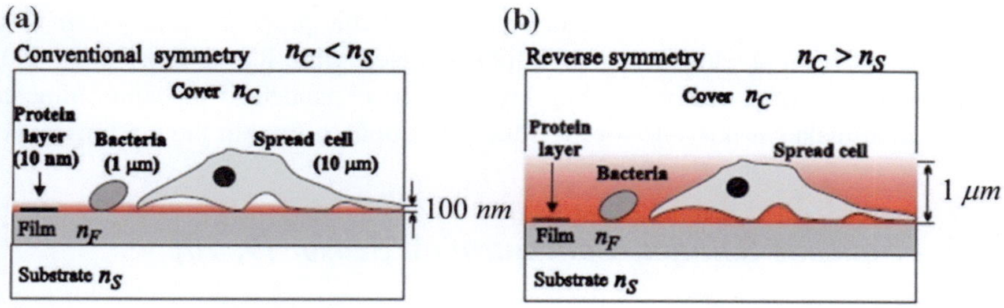

Fig. 8.12 Comparison of penetration depth between the conventional waveguide (**a**) and the reverse symmetry waveguide sensor (**b**)

reverse symmetry waveguide is very suitable for the detection of cells that are typically a few microns in size [7]. A comparison of penetration depth between conventional waveguide and the reverse symmetry waveguide sensor is shown in Fig. 8.12. The detection limit of RI is estimated to be 5×10^{-5} [8], a few dozen times better than that in the conventional waveguide.

8.4 Oscillating Wave Sensors Based on the Light Intensity

The common feature of the evanescent wave sensor is that the sample to be detected locates in the region where the evanescent wave of the resonant modes propagates. It has shown that the sensitivity of such sensors depends strongly on the power distribution of the resonance mode. The bigger the portion of power that propagates in the sample, the higher the sensitivity will be. To achieve a higher sensitivity, it is essential to get as much of the optical power as possible to propagate in the sensing region. Investigation of the mode power distribution suggests us to design a configuration that contains the sample in the guiding layer of the waveguide, where oscillating wave locates and most of the mode power concentrates.

However, this idea is blocked by two barriers in the conventional waveguide. Firstly, only when the RI requirement of $n_{\text{guiding}} > n_{\text{substrate}}, n_{\text{cladding}}$ is satisfied, a sandwiched structure can form a waveguide that light wave can propagate stably in the guiding layer. Since the RI of aqueous solutions is usually lower than the conventional waveguide materials, it is hard to find suitable material to fabricate the substrate and the cladding layer of the waveguide. Secondly, in the conventional waveguide structure, the thickness of guiding layer is usually around micrometer scale, and this limited sample room would make the sample into and out of sample room to be very difficult.

In the following sections, we elaborately describe our proposed oscillating wave sensors based on the SMCW structure. In SMCW, the above-mentioned problems are no longer obstacles. As the real part of dielectric constant of metal claddings is negative, it suggests that not only aqueous analyte but also gas can work as the guiding layer to be probed. So the detect range of this proposed sensor can vary from gas with RI as low as 1 to liquid samples with RI of more than 2.0. Furthermore, the thickness of guiding layer can be expanded to the submillimeter scale, which makes it possible for the aqueous sample to flow in the guiding layer.

8.4.1 Aqueous Solution Concentration Sensor [9, 10]

The configuration of oscillating wave sensor for the aqueous solution concentration sensing is shown in Fig. 8.13 [9]. A spacer O ring is sandwiched between two thin gold films coated on a coupling prism and a substrate glass plate, respectively, to form a sealed sample cell. The coupling prism is made of glass, with RI of 1.5. A gold film of 30 nm is deposited on its bottom face, with dielectric constants of $-11.4 + i1.5$ at the wavelength of 650 nm. The thickness of the spacer O ring is $h = 500$ μm and hence the same thickness for the sample layer. Another gold film with the thickness of 300 nm is deposited on a glass plate substrate. Aqueous sample is pumped through the inlet and the outlet hole on the glass plate into the sample cell by a peristaltic pump. In such a manner, an oscillating wave sensor is constructed, in which the sample sealed in the flow cell serves as the guiding layer of the SMCW.

After passing through a spatial filter, a collimated TE-polarized beam is focused by a lens and incident into the sensor structure, and its aperture angle is about 0.7°. The lens and the sensor structure are adjusted carefully to put the focal plane on the base of the coupling prism. At the synchronized angles, energy of the incident light is coupled into the waveguide structures to excite the ultrahigh-order modes. Several fringes will then be presented in the reflectivity pattern and captured by a

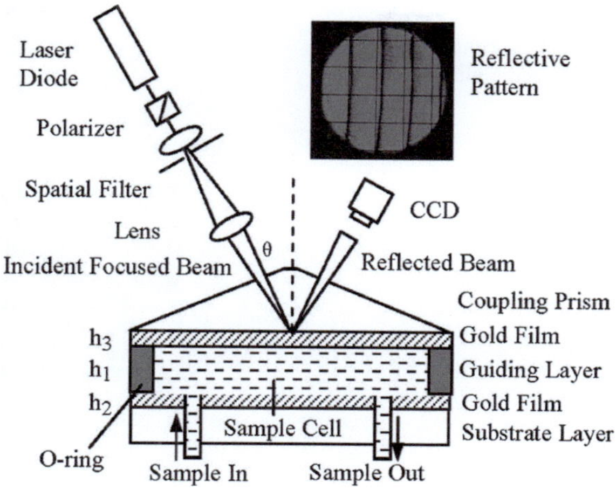

Fig. 8.13 Sensor structure and experimental layout

charge-coupled device (CCD) camera. A set of NaCl water solutions with concentration change of 250 ppm is used as sample analytes to be probed. As the concentration of analytes pumped in the flow cell is changed, the fringe position will shift due to the minute RI change. The experimental results show that a 250 ppm NaCl concentration change in the sample, corresponding to a change of 33×10^{-5} RIU (refractive index unit), will consequently result in a fringe shift of around 0.014°, as shown in Fig. 8.14. That means a sensitivity of 424°/RIU for TE

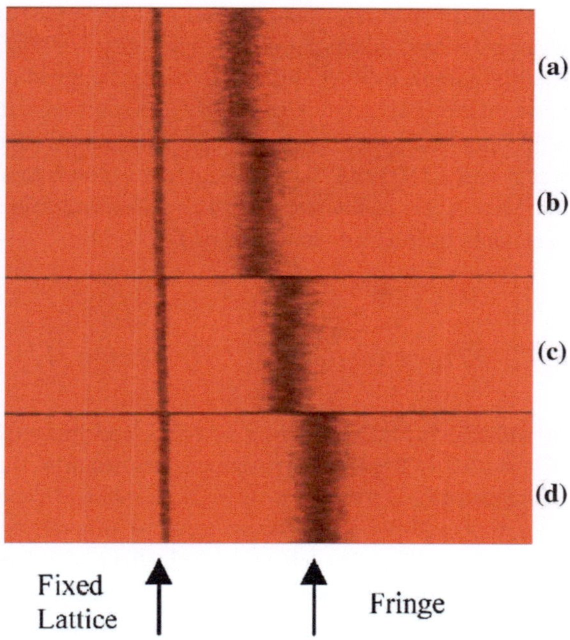

Fig. 8.14 Fringe shift captured by CCD, the lattice is in fixed positions for reference, the samples are **a** pure water, **b** 250 ppm NaCl solution, **c** 500 ppm NaCl solution, and **d** 750 ppm NaCl solution

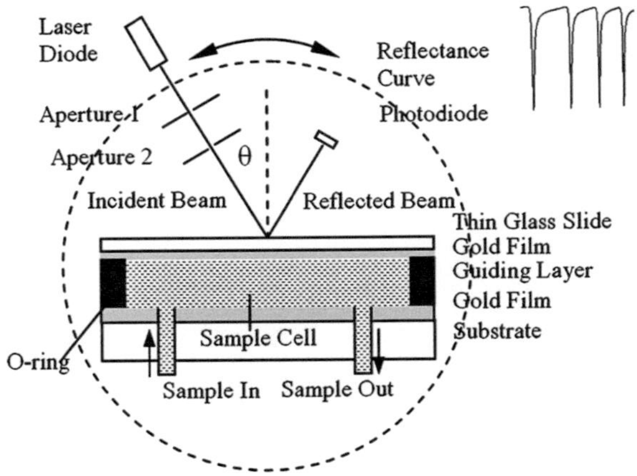

Fig. 8.15 The aqueous solution concentration sensor based on the intensity measurement

mode. For comparison, in the reserve symmetrical waveguide sensor presented by Horvath et al. [8], the experimental achieved sensitivities of TM and TE modes are 33.5°/RIU and 18.8°/RIU, respectively.

A minute variation in the concentration of the aqueous solution leads to a RI change of the guiding layer, which will cause an angular shift of the resonance dip. If the incident angle is fixed at the middle point of the falling or the rising edge, the reflected intensity is another desired variable to monitor the concentration of aqueous solution [10]. An aqueous solution concentration sensor based on the intensity modulation method is shown in Fig. 8.15. The detailed structure parameters can be found in [10].

The experimental sensor response is illustrated in Fig. 8.16, a 50 ppm NaCl water solution can cause a reflectance change of about 15 %, and a 150 ppm NaCl solution sample even causes a reflectance change of about 52 %. As the relationship between the concentration of NaCl water solution and its RI is known, it means that a 50 ppm NaCl water solution can bring an RI change of 6.6×10^{-6} over pure water, and a 150 ppm NaCl solution corresponds to a 2×10^{-5} RIU change. So the ideal resolution of the sensor is then calculated to be 8.8×10^{-8}, by assuming the detectable reflectance change to be 0.2 % [4].

8.4.2 Trace Chromium (VI) Sensor [11]

In addition to the weathering of rocks and the erosion of soils, trace chromium (Cr) is usually found in environment as a result of the discharge of many industrial manufacturing activities, such as stainless-steel production, leather tanning, electroplating, and pigment fabrication. Cr is generally in two most stable states of Cr (III) and Cr (VI). The physiological effect of these two states on the biological systems is totally opposite: Cr (III) is essential to human health at trace level, whereas Cr (VI) is readily absorbed by the lungs, digestive tracts, mucous

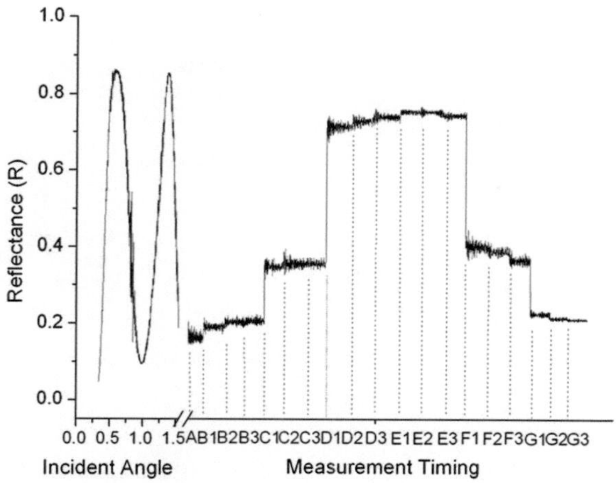

Fig. 8.16 Sensor response for NaCl water solution samples. The first intensity measurement for the pure water sample starts at *B1, B2,* and *B3* with 1-min intervals. 1 min after a 50 ppm NaCl solution is pumped in, a second set of measurement starts at *C1, C2,* and *C3* with 1-min intervals. Other timing spots are *D1, D2,* and *D3* for 150 ppm NaCl water solution; *E1, E2,* and *E3* for newly pumped in 150 ppm NaCl water solution; *F1, F2,* and *F3* for 50 ppm NaCl solutions again and back to purewater at *G1, G2,* and *G3*

membranes, and skins, is toxic and carcinogenic, and is considered as a serious pollutant in environment. The determination of Cr (VI) in environmental and at industrial sites is consequently important.

The minimum reflectivity of the ultrahigh-order mode is well described by Eq. (4.40) and can be cast in the form

$$R_{\min} \propto 1 - \frac{4\mathrm{Im}(\beta^0)\mathrm{Im}(\Delta\beta^L)}{\left[\mathrm{Im}(\beta^0) + \mathrm{Im}(\Delta\beta^L)\right]^2}, \tag{8.9}$$

where $\mathrm{Im}(\beta^0)$ and $\mathrm{Im}(\Delta\beta^L)$ represent the intrinsic damping and the radiative damping, respectively. The relationship between the concentration of solution and the imaginary part of dielectric constant can be derived from the Beer–Lambert law as below [12]

$$\mathrm{Im}(\varepsilon) = \frac{nK_m}{\ln 10 \cdot k_0} C, \tag{8.10}$$

where C is the concentration of solution, K_m is the molar absorptivity, and n is the real part of RI of the solution. As the concentration of Cr (VI) in aqueous solution changes, the extinction coefficient of the solution and the intrinsic damping of SMCW change too. Combining Eqs. (8.9) and (8.10), we can get the relationship between reflectivity and the concentration of solution.

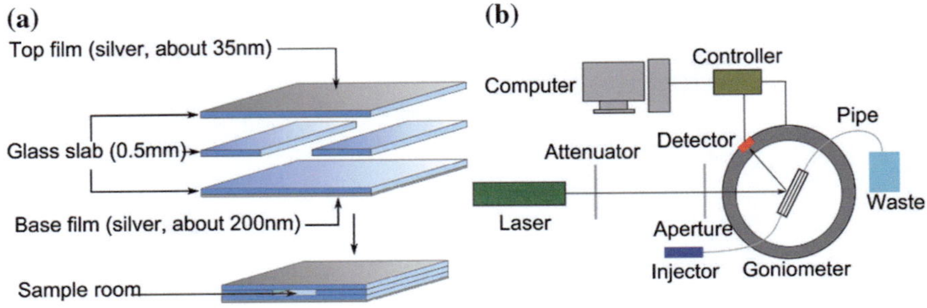

Fig. 8.17 The structure sensor (**a**) and experimental arrangement (**b**) for trace chromium (VI) sensing

As shown in Fig. 8.17a, the SMCW structure in the experiment is composed of three parts: (i) a polished glass substrate of 0.5 mm thickness has a thin silver film coated on the top side to act as a coupling layer, (ii) two parallel glass strips of 0.5 mm thickness are placed at a distance of 4 mm to form the sample confinement walls, (iii) another polished glass substrate of 0.5 mm thickness has a relatively thick silver film deposited on the base side to prevent light leakage. The experimental setup is shown in Fig. 8.17b. A TE-polarized laser beam from a solid-state laser is incident on the upper silver film at certain optical angles. An aperture with a diameter of 1 mm was inserted into the beam path to further confine the divergence angle to about 0.4 mrad. The sample solution was pumped into and out the cell room by a syringe through inlet and outlet pipes with 0.5 mm inner diameter. A computer-controlled $\theta/2\theta$ goniometer was used to carry out the angular scans. The intensity of the reflected beam was detected by a photodiode.

Determination of Cr (VI) was based on the color reaction with 1, 5-Diphenylcarbazide (DPC, C13H14N4O). The reaction takes place at a pH of 1.0 ± 0.3 and forms absorbing complex at the wavelength around 540 nm. Therefore, a laser of 532 nm was used as the light source in the experiment. Sulfuric acid (1:1) and phosphoric acid (1:1) were prepared by diluting 50 ml of acid in 50 ml of water. Chromogenic reagent was prepared by dissolving 2 g of DPC in 50 ml acetone and then diluting in water with a 100 ml measuring flask. Chromogenic reagent was prepared just before use. A 400 µg/l of Cr (VI) stock solution was prepared by dissolving 0.2829 ± 0.0001 g of potassium dichromate in water with a 250-ml measuring flask. The Cr (VI) working standard solutions were prepared by appropriate dilution by water. Color reaction took place by adding 0.25 ml of sulfuric acid (1:1), 0.25 ml of phosphoric acid (1:1), and 1 ml of chromogenic reagents to 25 ml of each Cr (VI) working standard solution. The reaction lasted for 7 min. The concentration of Cr (VI) working standard solution selected in this experiment was 0, 0.24, 0.48, 0.72, 0.96, 1.2, 1.44, 1.92, 2.16, and 2.64 µg/l. After reaction, the mixed solutions were injected into sensor chip. And then reflectivity spectrum of each solution was measured by rotating the

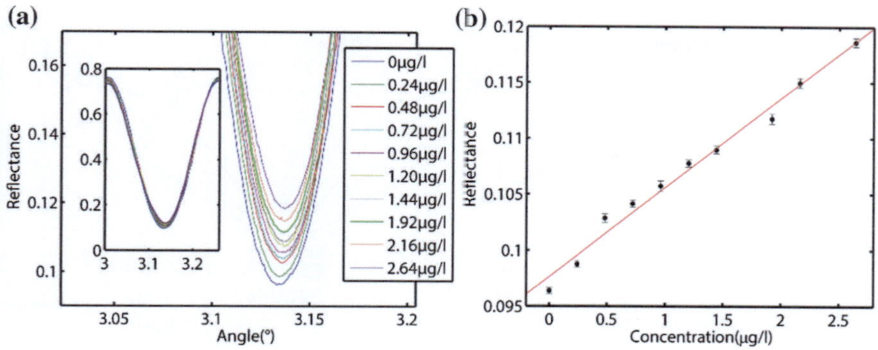

Fig. 8.18 The reflectivity spectrums (**a**) and the minimum reflectance (**b**) with respect to different concentration of Cr (VI)

goniometer. Before the injection of each solution, the sensor chip was washed by pumping 20 ml of deionized water to avoid residue.

Figure 8.18a shows the reflectivity spectrums of the sample solutions. As the Cr (VI) concentration increases, the reflectivity at coupled angle varies from 0.097 to 0.118. Figure 8.18b shows the relationship between minimum reflectivity and the concentration of Cr (VI). It produces an approximate linear response about Cr (VI) concentrations:

$$R_{\min} = (0.007965 \pm 0.000814) \cdot C(\mathrm{Cr\ (VI)}) + (0.09763 \pm 0.00117), \quad (8.11)$$

with the fitting relation coefficient, where C(Cr (VI)) is the concentration of Cr (VI) in μg/l. The limit of detection, defined as three times of the standard deviation of the measurement blank (a concentration of zero), is $\frac{3\times 0.00017}{0.007965} = 0.064\,\mu\mathrm{g}$ or 12 nM, which represents a 16-fold improvement compared to the surface plasmon field applied to the determination of Cr (VI) [13].

8.4.3 *Displacement Sensor [14–16]*

Displacement sensors have long been used in ultra precision measurements of geometrical quantities, such as positioning, vibration, and profile measurement of shafts and surfaces. As one of the commonly used displacement measurement techniques, interferometer method usually suffers from a relatively complicated optical arrangement and limited resolution. In this section, we describe our proposed displacement sensor based on the SMCW by analyzing the fringe shift [14] and by monitoring changes in the reflected intensity [15, 16]. These displacement sensors can work in real time and can provide accuracy better than those obtained by using photorefractive interferometer and speckle correlation technique.

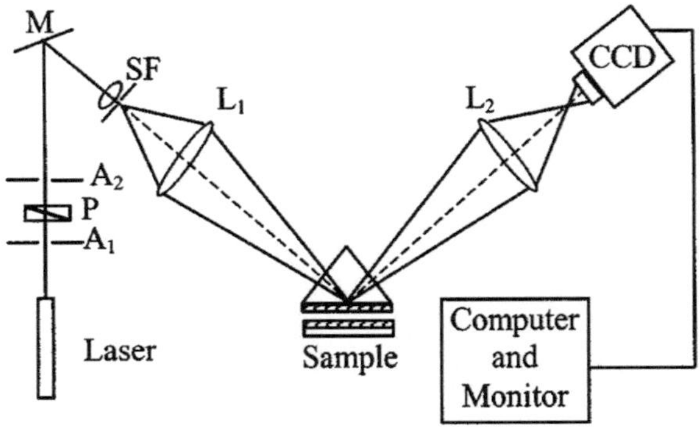

Fig. 8.19 Schematic arrangement of displacement sensing by analyzing the fringe shift

The experimental arrangement for the measurement of the displacement by analyzing the fringe shift is shown in Fig. 8.19. The SMCW consists of a stationary part 1 and a moving part 2. Part 1 is a glass prism, with a thin gold film of 50 nm deposited on its base, which is rigidly attached to a heavy bar in brass. Part 2, as an object of the measurement mounted on a translation stage, is a piece of flat glass coated with a relatively thick gold film of 200 nm. The thickness of the air gap is controlled by a differential micrometer fixed on the translation stage. A light beam from a laser passes through a Glan Thomson prism P and two apertures A1 and A2 and is then reflected by a mirror M. Being spatially filtered by a spatial filter, the input beam is finally focused by lens L1 at its rear focal plane, where the base of the prism is located. The reflective pattern, which is collected by another lens L2 and then recorded by a CCD camera, is placed a distance away from the focal plane of L2 along the optical axis. As a result of the energy transfer from the incident light into the guided mode, a fringe, which is attributed to the reduction in the intensity of the reflected bright spot, can be observed at the position of the resonance angle of the guided mode. If the moving part of the sample undergoes a slight displacement driven by the differential micrometer, the guide thickness will then be changed with the same amount of the displacement. Substantially, the fringe will shift its angular position as well. Using this known fringe shift quantity, the displacement of the object can be determined absolutely. The measurement range of our experiment is −12.5–240 μm, the accuracy of the displacement is about 50 nm, and the relative error is less than 0.4 %.

As shown in Fig. 8.20, the configuration of displacement sensor by monitoring the reflected intensity is composed of two parts: One is a glass prism on its base precoated with a thin gold film; the other is a 500-μm-thick $LiNbO_3$ slab sandwiched between two 400-nm-thick gold films and serves as a piezoelectric translator. The two components, separated by an air gap with a thickness of 100 μm, are rigidly mounted on a heavy platform. As applying a dc voltage on the pair electrodes of the piezoelectric translator, the air gap changes its thickness due to the piezoelectric effect of the $LiNbO_3$ slab. As a result, the reflection dip shifts its peak position and results in a change of the reflectivity. According to the resolution of the

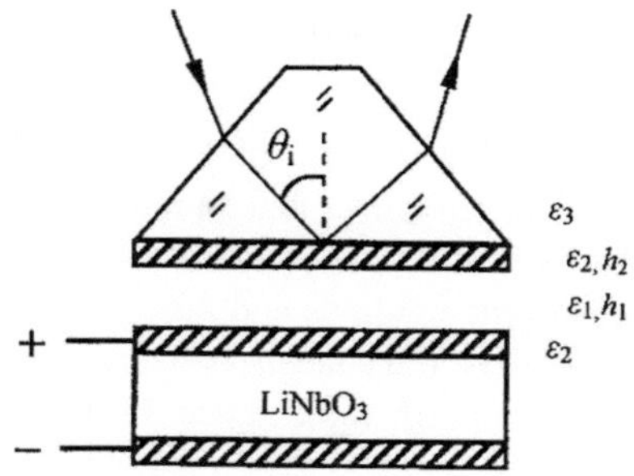

Fig. 8.20 Configuration of displacement sensor by monitoring the reflected intensity

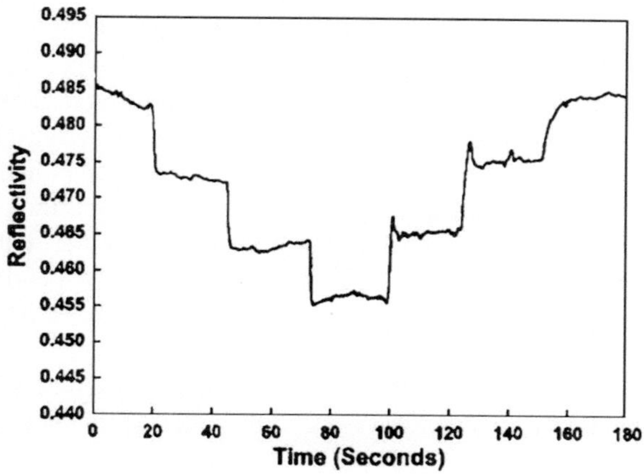

Fig. 8.21 Displacement sensitivity

reflectivity variation, displacement can be evaluated from the applied voltage and the piezoelectric coefficient of the $LiNbO_3$ slab.

The obtained displacement sensitivity is shown in Fig. 8.21. The waveguide thickness (h_1) is increased and decreased in steps by increasing and reducing the voltages applied on the electrodes of the piezoelectric translator. The step-style change of voltage is 50 V. According to the piezoelectric coefficient of a *Z*-cut $LiNbO_3$ slab, $d_{33} = 33.45\,\text{pm/V}$, the value of the displacement resolution for the proposed configuration is determined as $S = 50 \times 33.45 \times 10^{-3} = 1.7\,\text{nm}$, which corresponds to the reflectivity change of $\Delta R = 1\,\%$.

Subsequently, an improved scheme for displacement measurement based on intensity modulation and lock-in amplification techniques is proposed [16]. The noise can be restrained by the lock-in amplifier. From the experimental result, a displacement resolution of 3.3 pm is obtained.

8.4.4 Angular Displacement Sensor [17]

Angle displacement plays an important role not only in the alignment, assembly, and calibration of machine tools, but also in the fields of earthquake measurement and scientific experiments. There are two main ways to measure small angle: the electronic methods and the optical methods. As one of the most effective methods, capacitive transducers convert angular displacement to capacitance signal. By the technique of capacitance bridge, the method reaches the angle resolution of 1.0×10^{-7} rad. Optical methods have an excellent performance for measurement of angular displacement. Conventionally, the measurement for angle is based on the principle of the optical lever [18]. The method can amplify the angular information infinitely. However, it is extremely difficult to construct an angle sensor that incorporates both high resolution and compactness, and the stability of the angle sensor decreases as the size is increased to increase the resolution.

A layout of angular displacement sensor based on the SMCW is illustrated in Fig. 8.22. The device is composed of two components: a prism and an oscillator. The prism is attached tightly on a platform, and the oscillator, which is suspended separately by a thin tungsten fiber, runs parallel with the hypotenuse face of the prism. The oscillator freely rotates around its suspending point O as external torque is applied. There is an air gap with thickness h_1 between the prism and the oscillator.

Under action of the external torque, the oscillator rotates around the suspension point O with an angle of θ, and the variation in thickness of the air gap is expressed as

$$h_1 = h_{10} + l\theta, \tag{8.12}$$

where h_{10} is the thickness of the air gap in the absence of external torque, and l is the distance between the incident point of the laser beam and the suspending point O. In the intensity measurement scheme, a photodiode is employed to monitor the

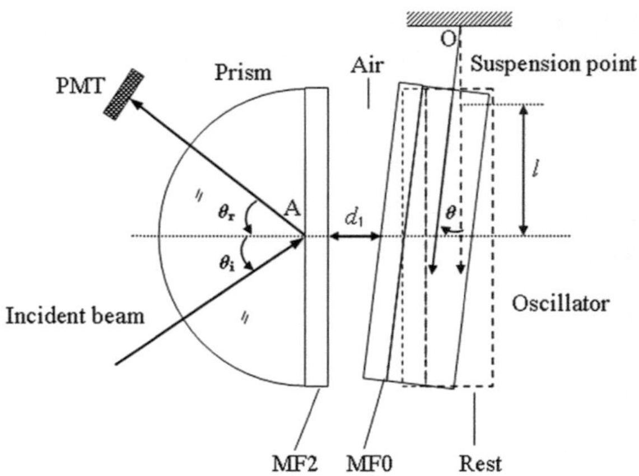

Fig. 8.22 Layout of angular displacement sensor based on the SMCW

variation in the optical intensity of reflected light. The minimum detectable angular displacement is 1.0×10^{-11} rad with a range of $\pm 2.5 \times 10^{-9}$ rad.

8.4.5 Wavelength Sensor [19]

Wavelength sensing, which can monitor and stabilize the spectrum of semiconductor diode lasers, is an important task in many fields of applications. In order to obtain high sensitivity, a probe with strong wavelength dispersion is often needed. Various probes such as arrayed-waveguide grating, chirped grating, three-layer slab waveguide, and the Fabry–Pérot etalon have been proposed. Among these methods, the guided mode of a three-layer slab waveguide structure is a preferable choice in wavelength sensing due to the high index contrast between the guiding layer and the cladding layer, which allows for strong optical confinement. However, the fact that a large part of the guided mode energy still disperses into the evanescent field decreases the sensitivity.

The ultrahigh-order modes excited in a SMCW have exhibited several attractive properties such as polarization-independent and strong wavelength dispersion [see Eq. (6.33)]. Experimental result is shown in Fig. 8.23. The required wavelength variation is achieved by temperature tuning of the tunable laser. The step-style change of wavelength is 0.5 pm, with the average 2.5 % change in the reflectivity ΔR. The sensitivity of about $5 \times 10^{10}\ \mathrm{m}^{-1}$ is obtained with a range of 18 pm, which covers the reflectivity from 0.1 to 0.7.

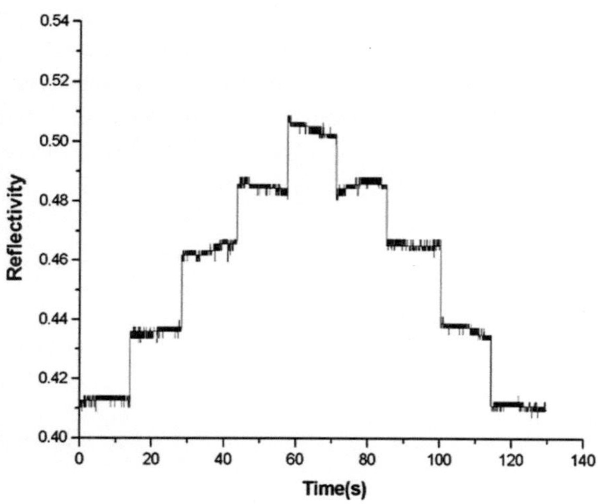

Fig. 8.23 The obtained wavelength sensitivity. The shift in the reflectivity curve was measured by changing the wavelength in steps of 0.5 pm

8.5 Oscillating Wave Sensors Based on the Goos–Hänchen Shift

The most popular sensing schemes in optical sensors are either monitoring the change in the reflectivity by fixing the excitation wavelength and the incident angle or monitoring the change in resonant wavelength or angle in response to the environmental change. Nevertheless, the best performance is demonstrated by using a phase-sensitive measurement. The reason is that the phase is shown to change far more rapidly as a function of environmental parameters compared to the change in reflectivity, wavelength, or angle. However, this improved performance always comes at a cost of the complexity in the optics. On the other hand, the Goos–Hänchen (GH) shift, which is a discrepancy of the reflected light point from its incident one, is found to be proportional to the first derivative of the reflected phase. As the environmental parameters can exert a significant influence on the GH shift, direct monitoring of the GH shift seems to be a promising way to build a simple and stable phase-sensitive detection scheme. In this section, we present various optical sensors, which can avoid the complicated fabrication and reduce disturbing from power fluctuation, by measuring the SMCW enhanced GH shift.

8.5.1 Aqueous Solution Concentration Sensor [20]

The schematic diagram of the aqueous solution concentration sensor is illustrated in Fig. 8.24. A glass prism is coated with a 20-nm-thick gold film to serve as the coupling layer. A 300-nm-thick gold film is sputtered on a glass slab to act as the substrate. The air gap of 0.7 mm sandwiched between the two gold films works as the guiding layer, where a gasket is used to form a sealed sample cell. With the help of a peristaltic pump, sample liquids to be detected flow into the cell through the inlet and the outlet tubes embedded in the substrate glass plate.

The theoretical sensitivity of the aqueous solution concentration sensor is defined by the change rate of the GH shift (L) with respect to the RI of the guiding layer (n_g), and it can be written as

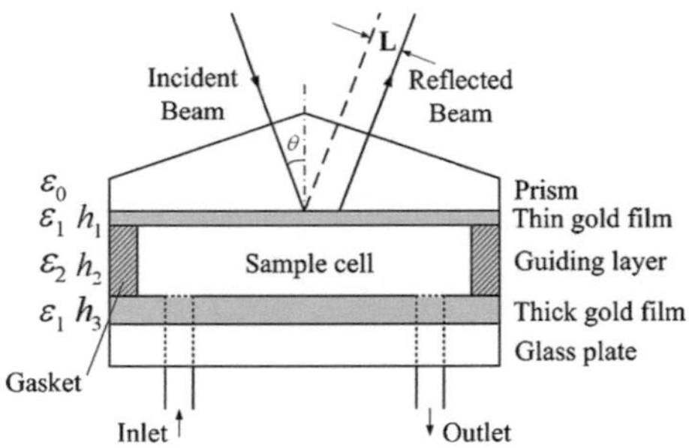

Fig. 8.24 Configuration of the aqueous solution concentration sensor

$$S = \frac{\mathrm{d}L}{\mathrm{d}n_g} = \left(\frac{\partial L}{\partial N}\right)\left(\frac{\partial N}{\partial n_g}\right) = S_1 S_2, \tag{8.13}$$

where N is the effective RI of the guided mode. A high sensitivity S_1 can be obtained when the incident angle gets close to the resonance peak. According to the dispersion equation for the ultrahigh-order modes in the SMCW, S_2 can be cast in the form

$$S_2 = \frac{n_g}{N}\frac{P_g}{P_T}, \tag{8.14}$$

where P_g and P_T represent the power flowing in the guiding layer and the whole waveguide structure, respectively. For the SMCW, the energy is almost totally coupled into the guiding layer ($P_g \approx P_T$) due to the strong confinement effect. Moreover, the allowed range of the effective refractive index is $0 < N < n_g$, which makes it possible to obtain an enormously high sensitivity ($S_2 \gg 1$), when the ultrahigh-order modes ($N \rightarrow 0$) at small incident angles are used as sensing probe.

In the experiment, the incident angle is adjusted to coincide with an ultrahigh-order mode. And this incident angle will be fixed to keep the stability of the optics. In order to attain a high sensitivity and a good linearity, the light wavelength is tuned by using a tunable laser. As the wavelength of the incident light changes, the GH shift of the reflected beam varies correspondingly. The operation wavelength is selected at the middle area of the falling edge of the resonance peak where a good linearity and a high sensitivity can be achieved. A tiny variation in the concentration of the sample solution, which corresponds to a RI change of the guiding layer, will lead to a change of the GH shift due to the translation of the resonance peak. The change of the GH shift is measured by using a position-sensitive detector (PSD). A series of NaCl solutions with the change step of 20 ppm in concentration is used as sample analyte to be probed. The experimental result is shown in Fig. 8.25. The step change of 20 ppm NaCl solution in concentration, which corresponds to a variation of 2.64×10^{-6} RIU, induces a GH shift change of at least 20 μm. By considering the noise level in the experiment, the probing sensitivity of 2.0×10^{-7} RIU is resolved since the measurement variation in the GH shift is confined within 1.5 μm for each sample.

8.5.2 Displacement Sensor [21]

In Sect. 8.4.3, the oscillating wave for displacement sensing with high resolution was developed by using SMCW structures. First, based on the fringe shift technique in the non-scanning configuration, a displacement resolution of 50 nm is demonstrated. Later, by monitoring the intensity of the reflected light, the displacement resolution achieves 1.7 nm, but the sensitivity of this sensor is still limited by the fluctuation of the laser source. In this section, an alternative approach based on the enhanced GH effect is presented to obtain a higher resolution and prevent the disturbance caused by the power fluctuation in the light source.

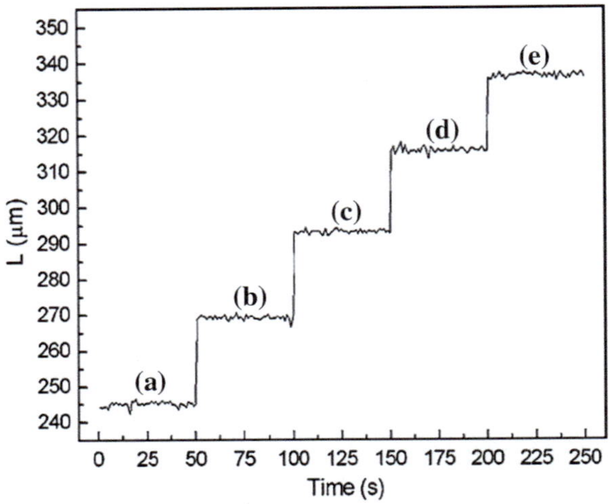

Fig. 8.25 The GH shifts with respect to the solutions of different concentrations: **a** pure water, **b** 20 ppm NaCl solution, **c** 40 ppm NaCl solution, **d** 60 ppm NaCl solution, and **e** 80 ppm NaCl solution

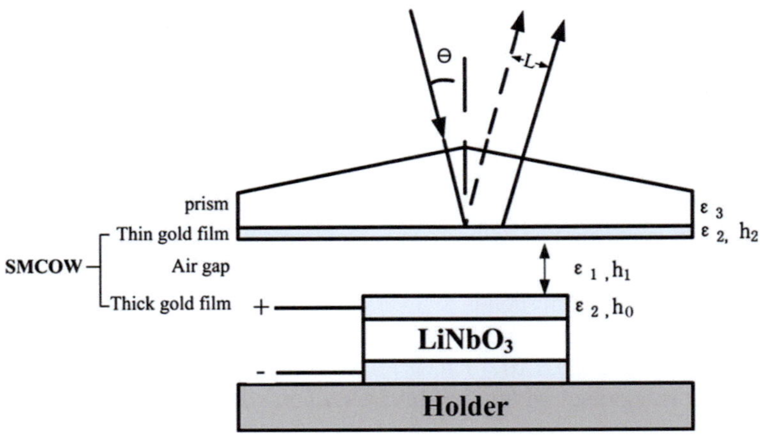

Fig. 8.26 Experimental structure of a displacement sensor based on the GH shift

Configuration for the displacement sensing is shown in Fig. 8.26. It consists of two parts: One is a glass prism with a thin gold film on its base; the other one is a z-cut $LiNbO_3$ slab sandwiched between two 300-nm-thick gold film electrodes and serves as a PZT. Two components separated by an air gap with a thickness of 500 μm are rigidly mounted onto a heavy platform to prevent the relative shift. While applying a dc voltage on the pair of electrodes of the piezoelectric translator, the air gap then changes its thickness owing to the piezoelectric effect of the $LiNbO_3$ slab.

The experimental result is shown in Fig. 8.27. The voltage applied on the piezoelectric translator between each step is 10 V, and the piezoelectric coefficient

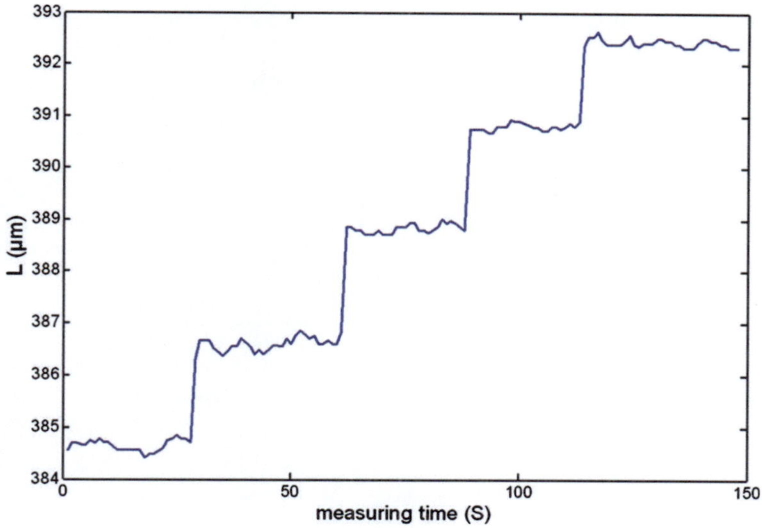

Fig. 8.27 Experimental sensitivity of the displacement sensor

of the z-cut $LiNbO_3$ is $d_{33} = 8 \times 10^{-12}\,\mathrm{m/V}$. Thus, the thickness change per step is determined as $\Delta d = 8 \times 10^{-12} \times 10\,\mathrm{m} = 8 \times 10^{-11}\,\mathrm{m}$, which leads to a GH shift change of 2 μm. The experimental ripple of each step is confined to 0.5 μm. With this noise level, the sensing resolution is evaluated to be 40 pm. Since the magnitude of the GH shift is irrelevant to the incident light intensity, a power fluctuation of the laser brings no disturbance to the resolution of the displacement sensor.

8.5.3 Wavelength Sensor [22]

In Sect. 8.4.5, an oscillating field for wavelength sensing based on the SMCW structure by measuring the reflectivity is described. Here, we present a wavelength sensor with a high resolution by monitoring the variation in GH shift of the reflected beam in the SMCW structure.

The schematic diagram of the wavelength sensor is shown in Fig. 8.28, which consists of two components. One is a glass prism with a gold thin film deposited on the bottom base. The other is a glass slab with another gold thin film coated on the upper surface. The mode dispersion for both TE and TM polarizations of the SMCW can be expressed as

$$\frac{\mathrm{d}N}{\mathrm{d}\lambda} = -\frac{\varepsilon_{\mathrm{air}} - N^2}{N\lambda}. \tag{8.15}$$

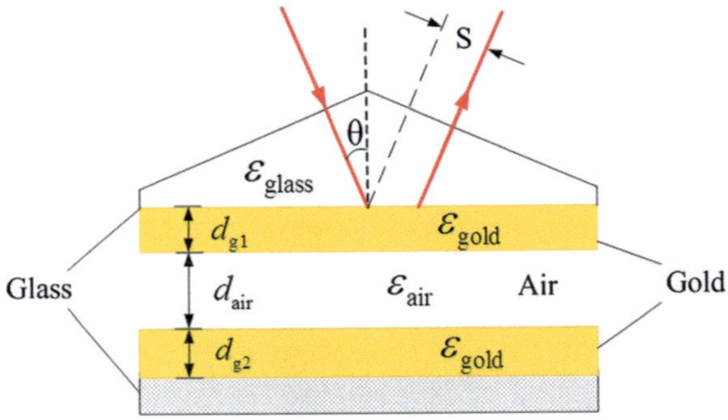

Fig. 8.28 Schematic diagram of the wavelength sensor

From Eq. (8.15), it is easy to obtain the wave vector dispersion as

$$\frac{d\kappa}{d\lambda} = 0, \tag{8.16}$$

for the normal component and

$$\frac{d\beta}{d\lambda} = -\frac{2\pi\varepsilon_{air}}{N\lambda^2}, \tag{8.17}$$

for the interfacial component. It is seen that κ retains invariant, while β changes rapidly with respect to the wavelength under the condition of $N \to 0$. The wave vector of the guided mode exhibits completely different dispersion properties in the two perpendicular directions, which indicates that the spatial position of the reflected beam will be highly sensitive to the light wavelength.

Wavelength-shift monitoring is performed with different thicknesses of the guiding layer in the SMCW, the results of which are shown in Fig. 8.29. Theoretical simulations based on the Gaussian beam model are also given for comparison.

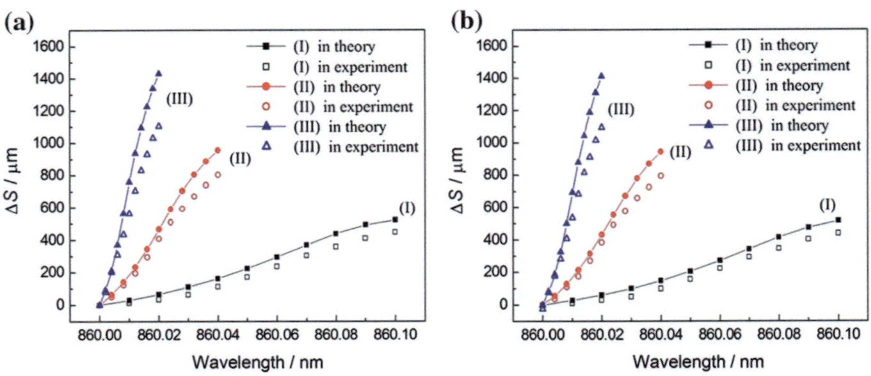

Fig. 8.29 Experimental result of GH shift variation (ΔS) with respect to the light wavelength. **a** TE polarization. **b** TM polarization. The datum points are of different parameters: *I* $d_{air} = 0.2\,\text{mm}$, $\theta = 3.72°$; *II* $d_{air} = 0.5\,\text{mm}$, $\theta = 3.84°$; and *III* $d_{air} = 1.0\,\text{mm}$, $\theta = 3.96°$

The sensitivity factor, which is defined as $F_S = \Delta S/\Delta\lambda$, reaches an average value of 4.5 μm/pm in the case of $d_{\text{air}} = 0.2\,\text{mm}$. The wavelength resolution of 0.33 pm is obtained, taking into account the noise level of GH shift measurement that is about 1.5 μm. The dynamic range of the wavelength-shift monitoring is 100 pm. As d_{air} increases, the probing sensitivity becomes higher correspondingly. The value of F_S is 20 μm/pm for $d_{\text{air}} = 0.5\,\text{mm}$ and 55 μm/pm for $d_{\text{air}} = 1.0\,\text{mm}$ in each spectral region, which denote a wavelength resolution of 0.075 and 0.027 pm, respectively. The dynamic range decreases to 40 pm for $d_{\text{air}} = 0.5\,\text{mm}$ and 20 pm for $d_{\text{air}} = 1.0\,\text{mm}$.

8.5.4 Enhanced Superprism Effect [23]

The superprism effect, which describes the unusual dispersion property much stronger than that in conventional prism, has attracted much attention since it promises extensive applications in optical communications, e.g., wavelength-division-multiplexing devices. Research has demonstrated that the anomalous dispersion could be obtained in one-dimensional (1D), two-dimensional, and three-dimensional photonic crystals (PCs) as well as multilayer thin-film structures. Instead of the limited effect based on material dispersion in conventional prisms, light of different wavelengths is enormously dispersed owing to the group-velocity dispersion and the phase-velocity dispersion near the bandgap of the PCs, which finally results in wavelength-dependent beam separation at the exiting facet. However, complicated fabrication and precise control were required in most of these devices that were constructed in quasi-periodic structures or photonic nanostructures.

As shown in Fig. 8.30, two gold thin films that are deposited on a prism base and a glass slab, respectively, by the sputtering technology, together with the air gap sandwiched between them, form the SMCW structure. The experimental result of the reflected beam displacement (S) is plotted as a function of the incident light wavelength. As depicted in Fig. 8.31, an extremely high spatial dispersion ability is exhibited on both the rising and falling sides of the displacement curve, where the GH shift rapidly changes with respect to the light wavelength. It is worth noting that

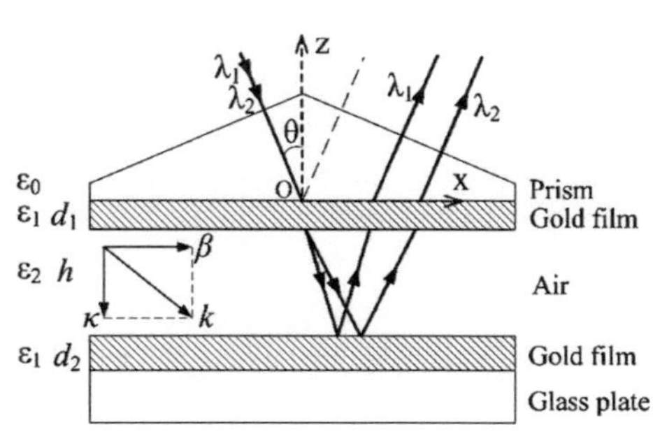

Fig. 8.30 Superprism effect in the SMCW structure

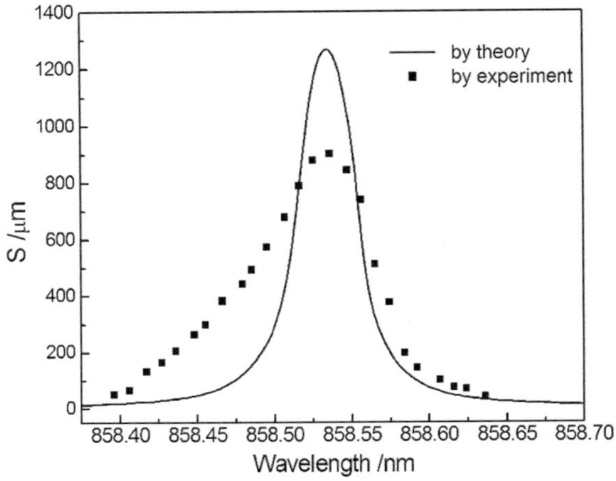

Fig. 8.31 Theoretical and experimental reflected beam displacements (*S*) with respect to the light wavelength

a maximum GH shift as large as 900 μm is obtained with the working wavelength interval ranging from 858.396 to 858.636 nm. In fact, the working region of wavelength can be tuned simply by changing the incident angle or the thickness of the guiding layer. Further experiments reveal that the deeper the penetration depth, the larger the GH shift becomes, namely the higher spatial dispersion ability will be. Through a tuning of 14 pm in wavelength, a lateral displacement as large as 1.5 mm is observed on the SMCW with a 3-mm-thick guiding layer [24].

8.6 Electro-optical Devices

8.6.1 Low-Voltage Electro-optic Polymer Modulator [25]

Electro-optic (EO) modulators are widely used in many photonic and optical/millimeter-wave systems. Taking advantages of the transverse Pockels effect, the transmission geometry can be arranged over a long interaction length to provide modulation at low voltages and high speeds, but this introduces subsequent alignment and propagation problems which lead to a large proportion of light energy loss. To decrease the propagation loss, much attention has been paid to EO modulators with reflection geometry. However, driving voltages for this type of devices are still too high to meet practical uses. For the sake of solving this problem, besides looking for materials, such as polymers, with much higher EO coefficients, it is found that resonant modes are highly sensitive to the electrically induced RI change in the guiding layer.

The schematic diagram of the low-voltage EO polymer modulator is shown in Fig. 8.32. It consists of a high-index cylindrical prism, a thin silver film as the coupling layer and also the top electrode, a poled polymer layer as the guiding layer, and another silver film as the base electrode. When a collimated

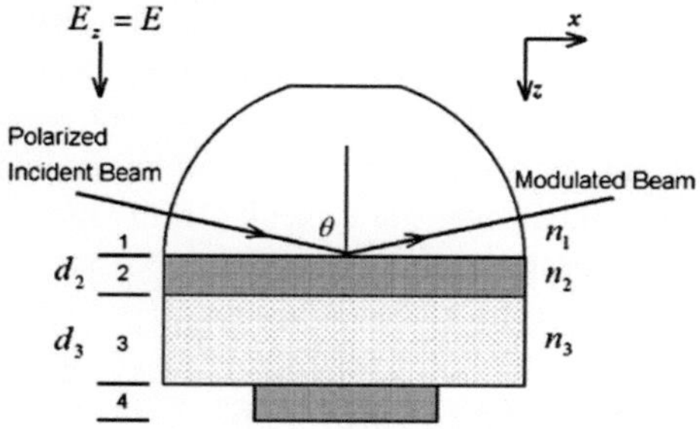

Fig. 8.32 Schematic diagram of the low-voltage EO polymer modulator: *1* prism; *2* silver film; *3* EO polymer film; and *4* silver film

Fig. 8.33 Modulated light reflectivity when RI of the EO polymer film varies as a result of applied voltage

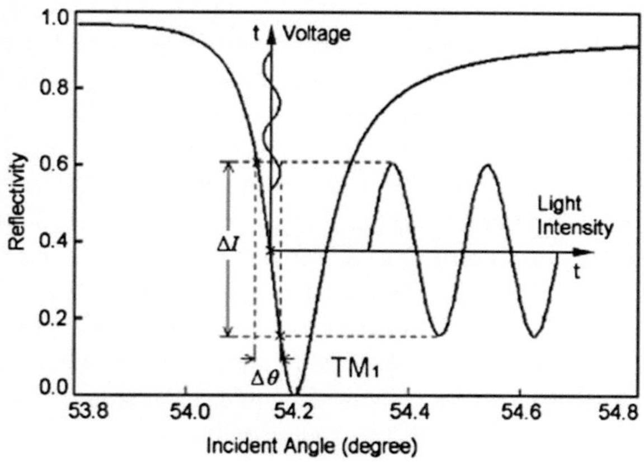

TM-polarized beam of light is incident upon the hypotenuse face of the prism, the electromagnetic field in the silver film beneath the prism is evanescent. At a synchronous incident angle, the evanescent field is phase matched to a guided wave in the polymer guiding layer, and the energy is coupled from the incident light into the guided wave. Applying an electrical field across the EO polymer film changes its RI and also changes the propagation constant of the guided wave, so the energy coupling efficiency and light reflectivity can be electrically modulated.

The process of the EO modulation is shown in Fig. 8.33. The modulation relationship can be written as

$$\Delta I = -\frac{\gamma_{33} k n_3^3 E}{2 n_1 \cos\theta}, \tag{8.18}$$

where ΔI is the light reflectivity that has been modulated, γ_{33} is the EO coefficient of the polymer, θ is the light incident angle, n_1 and n_3 are the refractive indices of the prism and the polymer film, respectively, E is the applied electrical field, and $k = \Delta I / \Delta\theta$ is the slope value of the fall-off in the resonance dip.

The measured modulation reflectivity was 60 % at a driving voltage of 12 V. This achievement is attributed to the extremely strong dependence of the guided mode on the electrically induced RI change in the EO polymer film. Since the propagation loss is absent, the total insertion loss of the modulator is below 1.5 dB, which is also an improvement over the typical 3–6 dB modulation loss in an EO modulator with transmission geometry.

8.6.2 Variable Optical Attenuator [26]

Variable optical attenuators are widely used in wavelength-division-multiplexing telecommunication systems to adjust power variations caused by changes in source power, amplifier gain, and other components. Commercially available variable optical attenuator devices are mainly based on optomechanical and TO effects and usually have response times of the order of milliseconds. With EO polymer, the response time is typically of the order of tens to hundreds of femtoseconds, but the intrinsic optical loss is usually higher than lithium niobate, resulting in a large proportion of light energy loss in the transmission geometry EO polymer device, such as the Mach–Zehnder interferometers. EO polymer device based on the reflection geometry does not present a serious loss problem due to the absence of propagation loss.

The diagram of the attenuator is similar to that of the EO polymer modulator shown in Fig. 8.32. Applying an electrical field across the EO polymer film changes its RI and also changes the propagation constant of the guided wave, so the energy coupling efficiency can be electrically controlled. Incident light can be totally absorbed or reflected at this angle, depending on the electrical field applied, so the device acts as an electrically controlled variable optical attenuator. Under the optimized gold film thickness, the dependence of the attenuation on applied voltage at different working interior angles is shown in Fig. 8.34. A maximum attenuation of 21.5 dB with the driving voltage of 33 V has been achieved. In order to further increase the attenuation range and decrease the working voltage, an EO polymer variable optical attenuator based on the cascaded attenuated-total-reflection technique is presented [27]. Results show that the dynamic range can reach 24.1 dB with an applied voltage as low as 39.6 V.

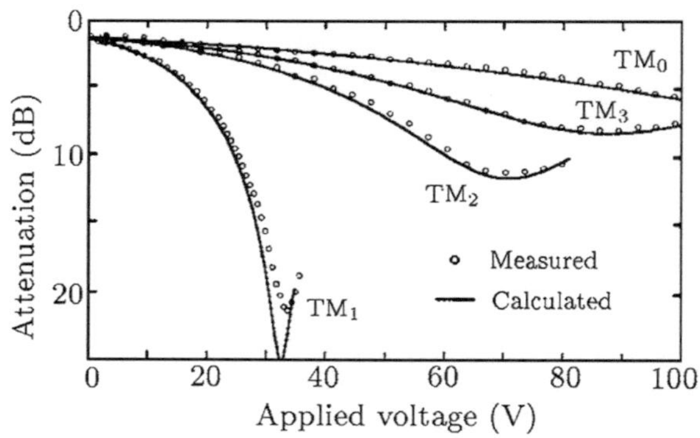

Fig. 8.34 Dependence of the attenuation on the applied voltage for different modes

8.6.3 *Tunable Polarization Beam Splitter [28]*

The technology of polarization beam splitter (PBS), which enables a spatial separation of the two orthogonal polarizations of light beam, can be widely used in optical systems, such as optical switches, data storage, and image processing. Different from conventional methods which rely on the inherent birefringence of anisotropic materials or the Brewster angle effect, various configurations for PBS which utilize different responses of the structure for TE and TM polarizations have been proposed, such as embedded metal-wire nanograting, coupled plasmonic waveguide arrays, and anisotropic metamaterial slab. However, since the functionalities of all these structures have been fixed at the fabrication, no further tuning of the optical properties can be performed, which is a limitation in many practical applications.

The nonlinear material $LiNbO_3$ is introduced in both the coupling prisms and the guiding layer of the SMCW, where the optical properties of polarization-dependent anisotropy and EO effect are put to use, respectively. As shown in Fig. 8.35, two prisms, which act as the coupling layer, are made of x-cut $LiNbO_3$ crystal. It is easily seen that TE- and TM-polarized beams are the extraordinary and ordinary waves inside the prisms, respectively. So polarization-dependent k_{y0} is obtained for the same light wavelength and incident angle owing to the different RIs of the prisms for the two orthogonal polarizations. And the z-cut $LiNbO_3$ slab, together with two air gaps, constitutes the guiding layers. The $LiNbO_3$ slab is the essential component for electrical tuning of the guiding layer properties, in which both TE- and TM-polarized beams can be considered as ordinary waves on condition of very small incident angles. The guiding layers are sandwiched between two gold thin films which are deposited on the bottom of the prisms. In the structure, the gold films not only serve as cladding layers but also work as electrodes, on which external electric field is applied.

As shown in Fig. 8.36, the spectral responses of the SMCW structure for both TE- and TM-polarized beams are calculated at the light wavelength of $\lambda = 860\,\text{nm}$.

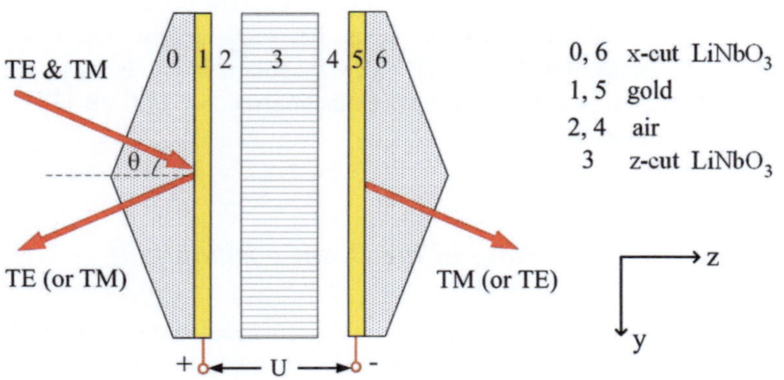

Fig. 8.35 Structure of the SMCW for tunable PBS

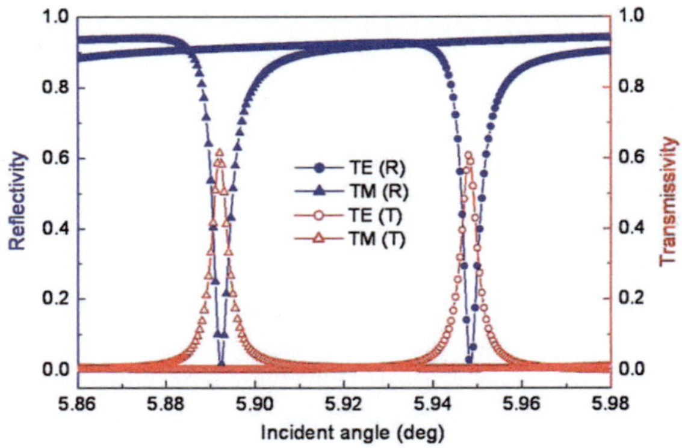

Fig. 8.36 Calculated reflectivity and transmissivity with respect to the incident angle for TE- and TM-polarized light

The reflectivity and transmissivity for each polarization strongly vary with the incident angle in the range of a guided mode. The coupling angles for TE and TM modes are staggered from each other, which give rise to different reflections and transmissions for different polarizations.

Initially, external voltage is not applied onto the structure, and the incident angle is settled at $\theta = 5.884°$. This angle is a little smaller than the resonance angle of TM mode, where the guided mode is not excited yet. The PBS effect is indistinct at this operation point since both TE and TM polarizations are highly reflected ($R_{TE} = 89.9\,\%$, $R_{TM} = 91.1\,\%$). Then, the applied electric field is adjusted to tune the PBS effect in the SMCW structure. As shown in Fig. 8.37a, the status of TE polarization remains nearly invariant with the augmentation of applied voltage, whereas TM polarization gradually changes into large transmission. The transmission of TM polarization reaches $T_{TM} = 60.3\,\%$ at $U = 2200\,\text{V}$, where a high separation of polarization beams is achieved. The extinction ratios in reflection and transmission are 16.9 and 15.7 dB, respectively. The situation of $\theta = 5.941°$, which is a little smaller than the resonance angle of TE mode, is also measured in the experiment (Fig. 8.37b). Similarly, high reflections for both TE and TM polarizations ($R_{TE} = 90.4\,\%$, $R_{TM} = 91.5\,\%$) are demonstrated at $U = 0$. But it presents an opposite way of PBS when adjusting the voltage, i.e., TM-polarized beam remains highly reflected, while TE-polarized beam becomes largely transmitted. The transmissivity (reflectivity) of TE (TM) polarization is $T_{TE} = 60.0\,\%$ ($R_{TM} = 90.8\,\%$) at $U = 2000\,\text{V}$, with an extinction ratio of 17.0 dB (18.6 dB) in TM (TE) polarization.

8.6.4 Electric Controlling of the Beam Position [29]

In this section, an optically nonlinear material is introduced into the guiding layer of the SMCW to control the GH shift of the reflected beam by applying an external electric field. As shown in Fig. 8.38, the configuration consists of two components.

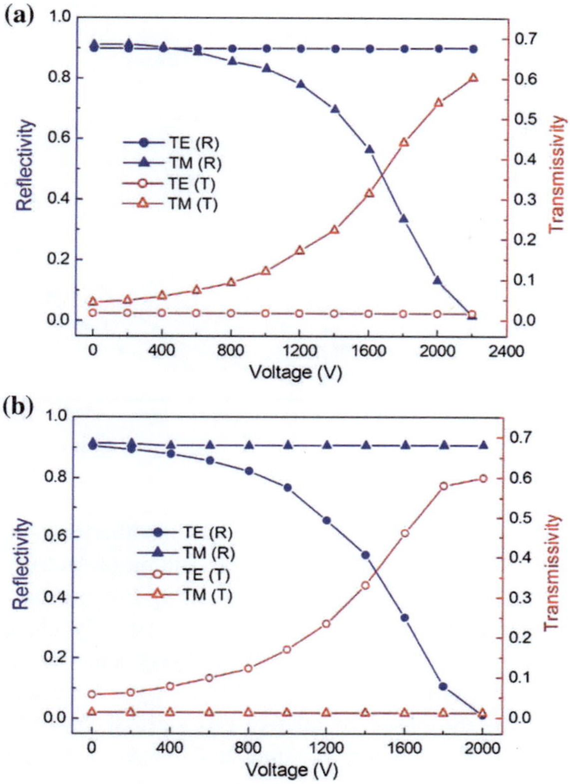

Fig. 8.37 Experimental measurements of tunable PBS. **a** $\theta = 5.884°$; **b** $\theta = 5.941°$

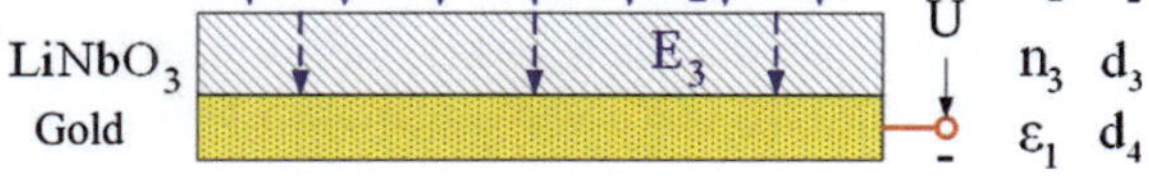

Fig. 8.38 The configuration for electric controlling of beam position

One is a glass prism with a thin gold film coated on the base. The other is a *z*-cut $LiNbO_3$ slab with another thin gold film deposited on the undersurface. Separated by an air gap, the two components are firmly emplaced onto a heavy platform to prevent relative moving.

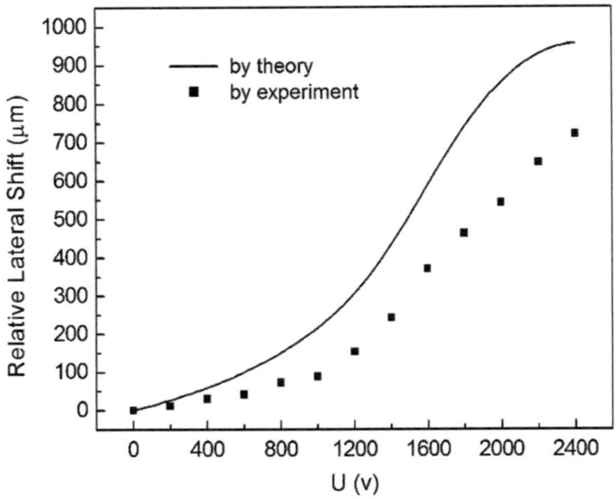

Fig. 8.39 GH shift of reflected beam with respect to applied voltage

When a dc voltage is applied on the two gold films, the RI n_3 and the thickness d_3 of the $LiNbO_3$ slab are changed due to the EO and piezoelectric effects. Besides, the thickness of the air gap d_2 is also changed during this process. The resonance condition of the ultrahigh-order mode will be dramatically changed due to the modification of the parameters of the guiding layers and finally give rise to a variation in the GH shift, which is monitored by a one-dimensional PSD. Figure 8.39 represents the observed relative GH shift of the reflected beam as a function of external applied voltage. To simulate the beam with a finite width (waist radius $w_0 = 800\ \mu m$), theoretical calculation based on the Gaussian beam model is also given for comparison. In the experiment, the relative GH shift of the reflected beam is continuously manipulated in a range of 720 μm by adjusting the voltage.

8.7 Research on Ferrofluid

8.7.1 Ferrofluid and Its Magneto-Optical Effects

Ferrofluid is a kind of stable colloids consisting of the magnetic nanoparticles (see Fig. 8.40) suspended in a carrier liquid. The most used nanoparticle material is magnetite (Fe_3O_4), and its average diameter ranges from 5 to 12 nm. The carrier liquids such as water, kerosene, or various oils are available. With the aid of coated surfactants, these nanoparticles neither form sediment in the gravitational field nor agglomerate together. As a result, each nanoparticle behaves as a constant magnetic dipole moment and can be freely aligned by the affect of external magnetic fields. The magnetically induced structures in ferrofluid and the related magneto-optical effects have recently attracted the interests of many scientists due to their potential applications.

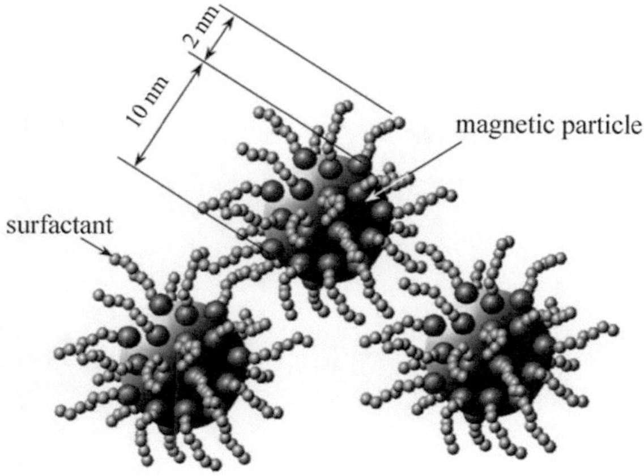

Fig. 8.40 The schematic view of the surfactant coated magnetic nanoparticles

To research the structure evolution, ferrofluid is usually sealed in a glass cell with a thickness of several to tens of micrometers. When a parallel magnetic field is applied, a portion of nanoparticles will agglomerate to form the needle-like chains. If the field strength is further increased, more nanoparticles will participate in the agglomeration to generate more needle-like chains. Meanwhile, two former separated chains will merge together to produce a longer chain. As shown in Fig. 8.41, a nearly one-dimensional periodic long-chain structure is obtained under the external parallel magnetic fields [30, 31]. Furthermore, it was observed that the spacing

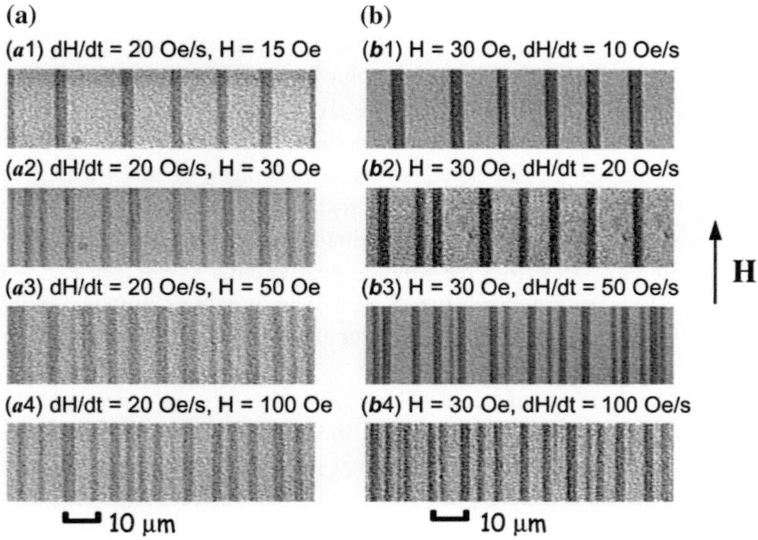

Fig. 8.41 Periodic long-chain structural patterns formed in a ferrofluid film subjected to parallel fields at: **a** different field strength with constant sweep rate and **b** different sweep rate with a given field strength

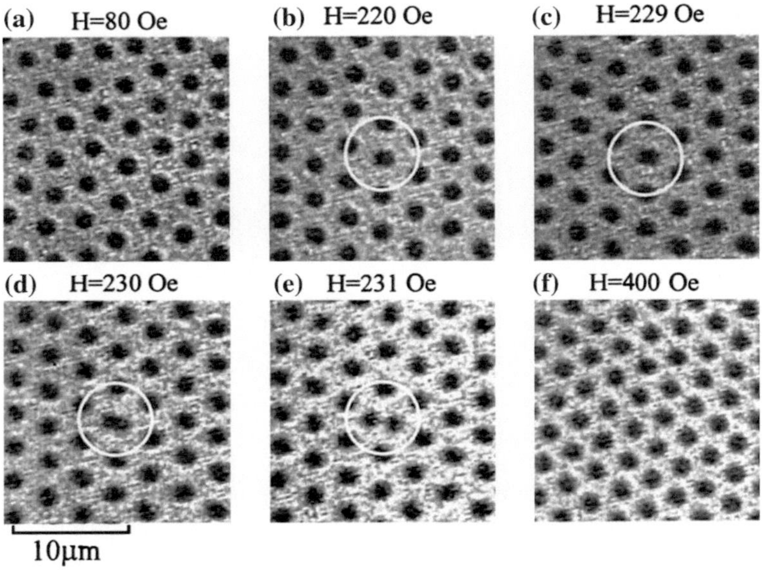

Fig. 8.42 Magnified images of a typical column breaking process in a ferrofluid film subjected to a perpendicular magnetic field

s between two neighboring long chains and the averaged chain width *w* decreased as the magnetic field strength *H* is increased.

In contrast to the needle-like chains, the cylindrical columns are formed in a ferrofluid film under a perpendicular magnetic field [32]. With an increase in the magnetic field strength *H*, three stages of the ferrofluid structure are subsequently evolved. In the beginning, the number of columns increases gradually. Columns are located randomly in the film, and their size shows a broadened distribution. Next, the ferrofluid structure evolves from a monodispersed state to an ordered one (see Fig. 8.42a), which can be characterized by the averaged distance *d* between two neighboring columns and the averaged diameter *a* of columns. The *d* is found to decrease under a height *H*, whereas *a* remained almost constant. During the finally stage, a column elongates from the original near circular shape (see Fig. 8.42b–d) and splits into two columns (see Fig. 8.42e). This process continues until all the columns split, and then, another order structure with a smaller *a* is formed (see Fig. 8.42f).

Due to the occurrence of needle-like chains, the ferrofluid becomes optically anisotropic and birefringent. Consequently, the transmission axis of the transmitted light rotates by an angle with respect to that of a linearly polarized light passing normally through the ferrofluid film [33]. Moreover, the transmittance and its response time are also magnetic field dependent. Under a perpendicular magnetic field, the resulted cylindrical column structure can diffract the visible light like an optical grating [34]. Figure 8.43 is a typical chromatic ring that resulted from the passage of a white light through the ferrofluid film. The spatial distribution of colors

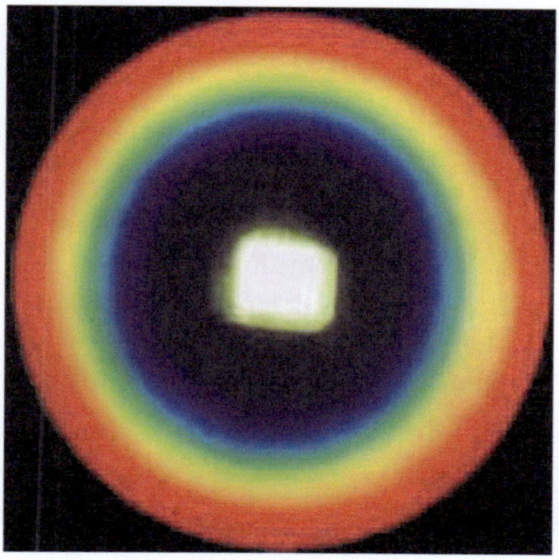

Fig. 8.43 Chromatic ring resulting from the diffraction as the parallel beam of white light passing through the ferrofluid film subjected to a perpendicular magnetic field

in the chromatic ring depends on the wavelength of light; red appears on the outmost ring, and, successively, orange, yellow, green, blue, and violet. These versatile magneto-optical effects of the ferrofluid have found broad applications in various functional optical devices, such as sensors, switches, modulators, and photonic crystal fiber [35, 36].

8.7.2 *Optical Trapping and Soret Effect*

Optically based manipulation techniques are playing an increasingly important role in the operation of microfluid. For instance, in the test of optical trapping technique [37], when a laser beam is tightly focused with an objective lens of high numerical aperture (NA), the dielectric particle near the focus will receive a force, which can be decomposed into two components: (1) a scattering force, in the direction of light propagation and (2) a gradient force, in the direction of the spatial light gradient. The reasons for such two force components are given as follows. Incident light impinges on the particle from one direction, but is scattered in a variety of directions. As a result, there is a net momentum transfer to the particle from the incident photons and the particle experiences a scattering force. On the other hand, the gradient force arises from the fact that a dipole in an inhomogeneous electric field will undergo a force in the direction of the field gradient.

When the trapped sphere is much smaller than the wavelength of the trapping laser, i.e., $a \ll \lambda$, the conditions for Raleigh scattering are satisfied and the optical forces can be calculated by treating the particle as a point dipole. For a sphere of radius a, the scattering component force is written as

$$F_{\text{scatt}} = \frac{I_0 \sigma n_m}{c}, \tag{8.19}$$

$$\sigma = \frac{128\pi^5 a^6}{3\lambda^4}\left(\frac{m^2 - 1}{m^2 + 2}\right)^2, \tag{8.20}$$

where I_0 is the intensity of incident light, σ is the scattering cross section of the sphere, n_m is the index of refraction of the medium, c is the speed of light in vacuum, m is the ratio of the index of refraction of the particle to the index of the medium, and λ is the wavelength of the trapping laser. The gradient force can be expressed as

$$F_{\text{grad}} = \frac{2\pi\alpha}{cn_m^2}\nabla I_0, \tag{8.21}$$

where

$$\alpha = n_m^2 a^3 \left(\frac{m^2 - 1}{m^2 + 2}\right) \tag{8.22}$$

is the polarizability of the sphere. It shows that the gradient force is proportional to the intensity gradient and points up the gradient when $m > 1$.

Recently, an experiment of using a visible light beam to optically trap the magnetic nanoparticles and thus to optically switch another near-infrared light beam was demonstrated [38]. The configuration of optical switch is shown in Fig. 8.44a. A capillary filled with a water-based ferrofluid is aligned in the x direction. A visible light is used as the trapping light and focused onto the capillary along the z direction by using an objective lens. Two optical fibers aligned in the y direction are used to emit and collect the near-infrared signal light. The switching mechanism is illustrated in Fig. 8.44b, c. In the absence of the trapping light (Fig. 8.44b), the signal light can be transmitted through the ferrofluid, because its wavelength is much larger than the diameter of the magnetic nanoparticles and the scattering is quite weak. Once the trapping light is imposed, however, the signal light is strongly reflected because of multiple scattering off laser-induced magnetic clusters, as shown in Fig. 8.44c.

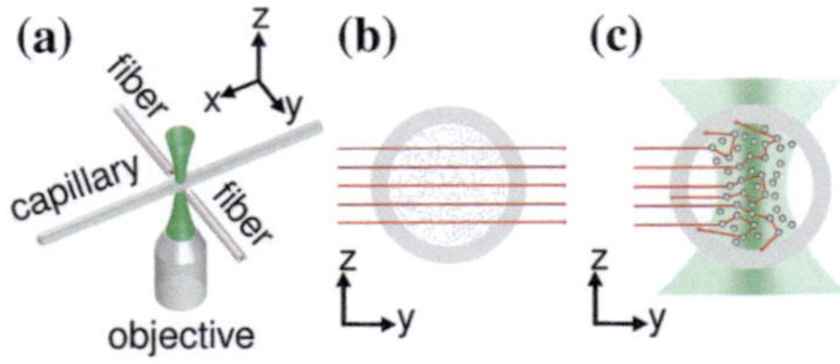

Fig. 8.44 Schematics showing **a** the structure of the proposed optical switch, **b** random distribution of magnetic nanoparticles inside the capillary and the transmission of the signal light through the capillary in the absence of the control light, and **c** the formation of magnetic clusters and enhanced backscattering of the signal light induced by the control light

The Soret effect [39] refers to the particle concentration variations due to laser beam-induced thermal diffusion in liquid mixtures. At the steady state, the relationship between the temperature and the concentration gradients is called as Soret constant S_T. In the case $S_T < 0$, the local increase of temperature attracts the particles into the warmer region. This yields increased light absorption, which further increases the temperature. Such a positive feedback may lead to an avalanche jump in the particle concentration and bistability. On the other side, the positive Soret constant can push the particles escaping from the hot regions to the cooler parts, thus decreasing the local temperature.

8.7.3 Magneto-optical Modulation

As described in Sect. 8.7.1, the transmittance of ferrofluid film is magnetic field dependent. However, due to limited experimental accuracy, few works addressed the issue of extremely dilute ferrofluids under weak field experimentally. By injecting the ferrofluid into a SMCW structure, even a slight perturbation in the formation structure of the nanoparticles will lead to a significant change in the reflection intensity [40]. To make this point clear, a simple three-layer optical waveguide consisting of a thin metal-coupling layer, a guiding layer, and a metal substrate is analyzed. Under the approximate phase-matching condition (Eq. 4.35), the reflectivity of the SMCW structure approaches its minimum and can be expressed by Eq. (4.40). The half width of the resonance dip, which is completely determined by the intrinsic and radiative dampings, is roughly approximated as

$$W_\theta = \frac{2\left[\mathrm{Im}\left(\beta^0\right) + \mathrm{Im}\left(\Delta\beta^{\mathrm{rad}}\right)\right]}{k_0\sqrt{\varepsilon_1}\cos\theta}. \tag{8.23}$$

It is easy to deduce the following relation from the dispersion equation of the guided modes

$$\mathrm{d}\theta = \frac{1}{\sqrt{\varepsilon_1}\cos\theta}\mathrm{d}N \approx \frac{n_3}{\varepsilon_1 \sin\theta\cos\theta}\mathrm{d}n_3. \tag{8.24}$$

So a rather rough estimation of the intensity variation in the reflected beam can be written as

$$\Delta R \approx \frac{1 - R_{\min}}{W_\theta}\frac{\mathrm{d}\theta}{\mathrm{d}n_3}\Delta n_3. \tag{8.25}$$

The subscripts 1 and 3 represent the free space and guiding layer, respectively (Fig. 8.45).

A continuous wave laser was used as a light source. In order to obtain a well-collimated, polarized laser beam, a polarizer and two apertures with a diameter

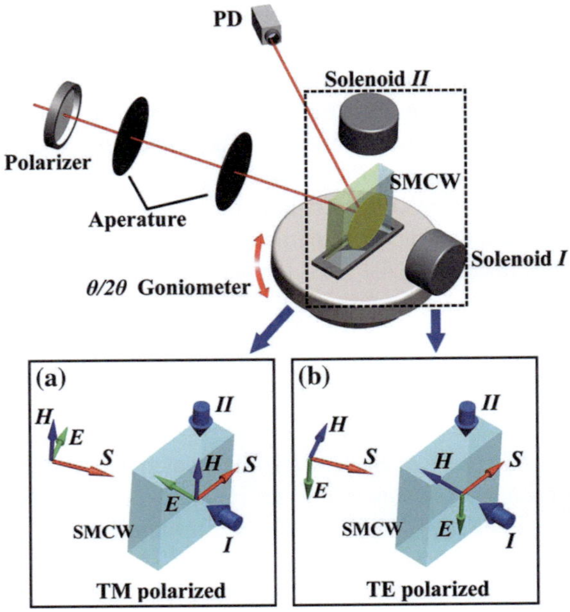

Fig. 8.45 Schematic of the experimental setup. The details at the *bottom* show the Poynting vector and electric displacement of the excited modes when the incident light is TE/TM polarized

of 1 mm were set between the laser and the SMCW structure. A solenoid that can be placed at fixed positions marked as "I" or "II" was used to provide a uniform magnetic field either perpendicular or parallel to the surface of the SMCW structure. The magnetic field was measured by a Gauss meter. Figure 8.56 illustrates the experiment setup, and its inset gives the Poynting vector and electric displacement of the incident laser and the guided modes, respectively, when the incident light is TM/TE polarized. The electric displacement vector of the TM-polarized incident light will be perpendicular to the structure surface, while the electric displacement vector of the TE-polarized light will be parallel to the structure surface. For simplicity, we will refer to the rays whose electric displacement vector is perpendicular/parallel to the magnetic field as ordinary/extraordinary rays.

From the above discussion, different choices of the polarization of incident light and the direction of the magnetic field will result in a total of four cases. When an external magnetic field is applied to the ferrofluids, a variation in the RI that results from the agglomeration of magnetic nanoparticles will occur. The concentration of the measured ferrofluids was 0.053 %; the incident angle was fixed at the middle area of the falling edge; and the maximum magnetic field strength used was only 24 Oe. The reflectivity variations are measured as a function of the magnetic field strength corresponding to all four cases, and the dependences of the normalized reflectivity on the field strength are plotted in Fig. 8.46. The magnetic field-dependent reflectivity shows different trends for ordinary and extraordinary rays. The reason is that the RI of the extraordinary (ordinary) ray increases (decreases) upon the application of the external field, no matter whether the magnetic field is perpendicular or parallel to the SMCW structure. The difference between the top

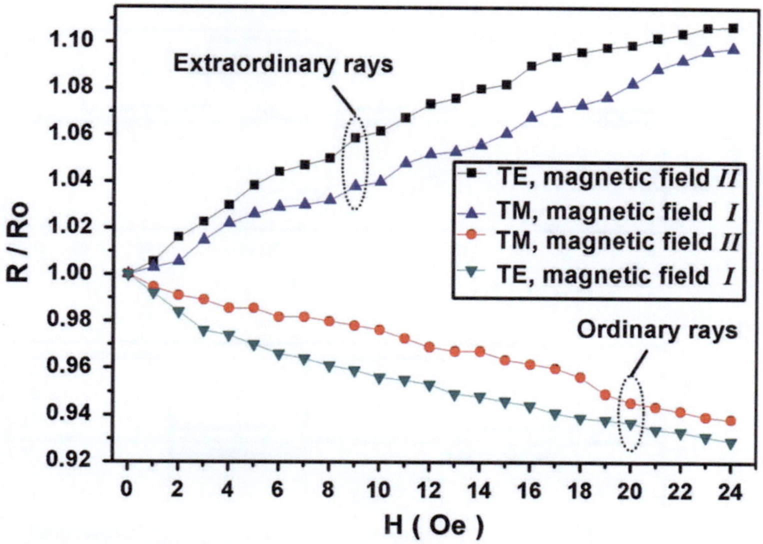

Fig. 8.46 Normalized reflectivity as a function of magnetic field for four different cases. The concentration of the ferrofluids measured was 0.053 %

(bottom) two curves in Fig. 8.46 probably rises from the difference in the basic Fresnel formula governing the reflection characteristic of multilayer planar waveguide for TE (TM)-polarized light.

Besides the transmission, the response time of ferrofluid film is also magnetic field dependent. A number of reports have pointed out that the agglomeration rate of magnetic nanoparticles is usually in second scale [41] or even in minute scale [42]. Furthermore, it is demonstrated that a finite retardation exists between the switch on/off of the external magnetic field and the variation in the transmission through the ferrofluid films, and this corresponding retarding time ranges from several to tens of milliseconds [43]. Our experimental results demonstrate that a switching time of a ferrofluid-sealed SMCW can be shortened to less than 2 ms [44].

A uniform magnetic field is applied parallel to the metal film surface of SMCW structure and perpendicular to the incident light beam. In Fig. 8.47, the time-dependent reflectivity is normalized with respect to that obtained without the magnetic field R_0. It was found that the reflectivity decreases quickly to a lower level in less than 1.2 ms when the magnetic field is switched on and rises to R_0 in less than 0.9 ms when the magnetic field is switched off. Figure 8.47b shows that when the modulation period of the magnetic field is prolonged to 4 s, the reflectivity remains nearly unchanged until the magnetic field is withdrawn. Based on the experimental results, the corresponding switching time is reduced by three orders of magnitude shorter than that reported in the literature [41]. However, in Fig. 8.47c, the switching times increase dramatically as the source is replaced by a confocal laser beam.

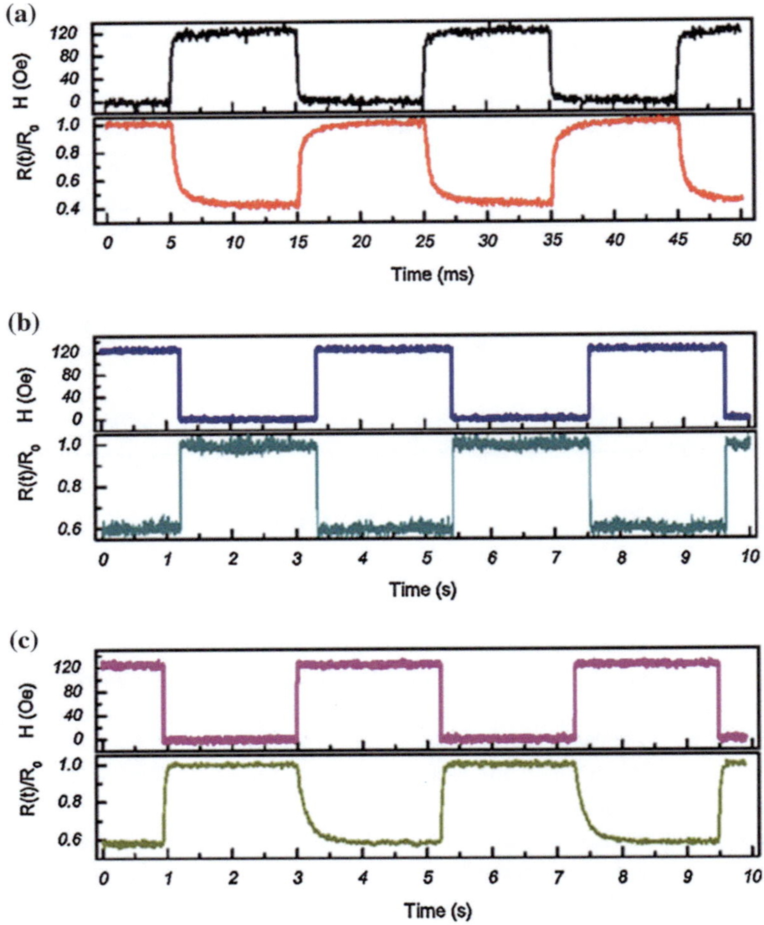

Fig. 8.47 Periodically applied magnetic fields and the corresponding dynamic variations in the normalized reflectivity. **a** The modulation period of the magnetic field is 20 ms. **b** The modulation period of the magnetic field is 4 s. **c** The modulation period of the magnetic field is 4 s, and a confocal laser beam is applied instead of the parallel laser beam

The experiments to investigate the intensity variations in the reflected beam upon simply turning on/off the laser source were also carried out. As shown in Fig. 8.48, the laser was turned on and off twice, while the incident angle is fixed at the middle of a falling edge without any external magnetic field. It can be seen that (1) there is an increase ($\Delta R = R_{\text{off}} - R_{\text{on}}$) in the reflectivity when the laser was switched on, and (2) the reflectivity gradually declines to the stable level R_{on} within about 7 s. Similar phenomenon can also be observed at the rising edge, and all these processes can be stably repeated. As shown in Fig. 8.49, the time-dependent reflectivity also exhibits the retarded effect upon the switching on/off of the external field. The

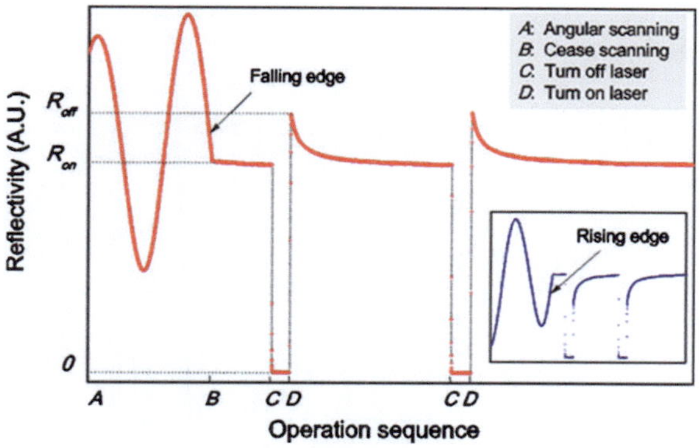

Fig. 8.48 The reflectivity sequence obtained by angular scanning and switching on and off the laser source. The incident angle is at the middle of a falling edge (*inset* the incident angle is at the middle of a rising edge)

Fig. 8.49 Time dependencies of the applied magnetic pulse and the normalized reflectivity. The operation incident angle is selected at a falling edge

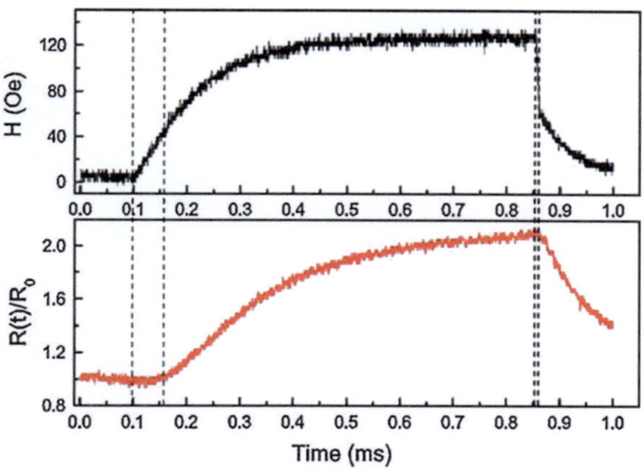

retarding time to the switching-on of the magnetic field is about 0.06 ms, while the retarding time to the switching-off of the magnetic field is about 0.01 ms.

To understand the physics behind the above phenomenon, the time-averaged *z*-component of the Poynting vector, S_z, of a pure ultrahigh-order mode is calculated (see Fig. 8.50). The amplitude of S_z in the guiding layer is enhanced by 68 times than that in the air before coupling into the SMCW structure. Figure 8.51 shows the electromagnetic energy distribution for an ultrahigh-order mode which oscillates periodically in space along *x*-axis. Besides, the excited ultrahigh-order mode propagates in *z* direction in the guiding layer, leading to the periodic distributed electromagnetic energy along the *z*-axis. Thus, in the *x*–*z* plane, the mode field pattern can be simplified as a checkerboard-like optical lattice. Due to the inhomogeneity of the electromagnetic energy, the magnetic nanoparticles within the excited mode field will all be optically trapped. As shown in Fig. 8.51, the

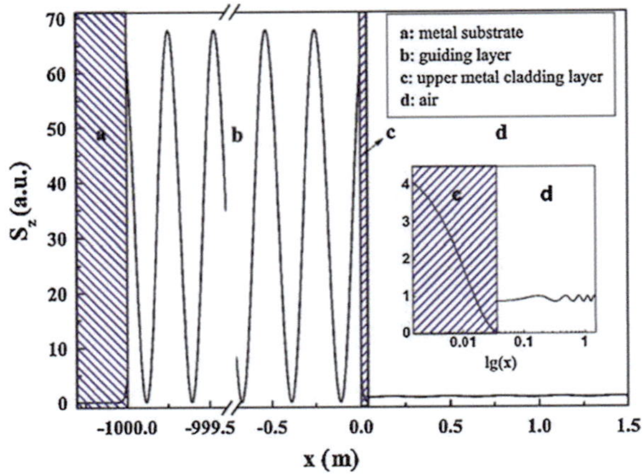

Fig. 8.50 Plot of z-component of the Poynting vector of a pure ultrahigh-order mode, and the maximum of the Poynting vector in air is normalized to unit

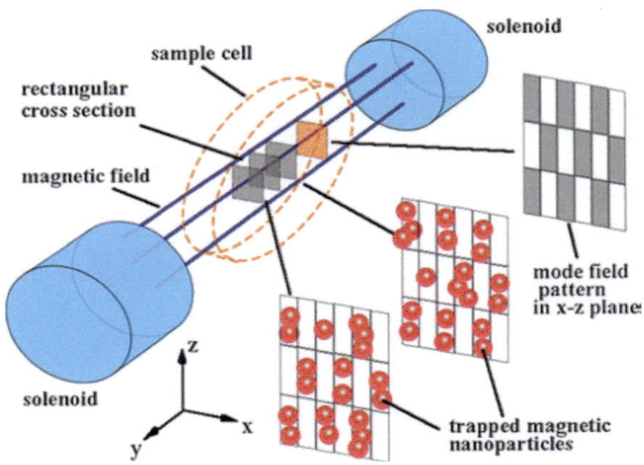

Fig. 8.51 Plot showing the optically trapped magnetic nanoparticles in the sample room

nanoparticles are trapped two dimensionally in x–z plane and distributed homogeneously along the y direction; thus, the column-like structure of nanoparticles parallel to the magnetic field has already been formed before the magnetic field is switched on. Consequently, when the external magnetic field is switched on, these nanoparticles in the sample room simply need to rotate to align their magnetic moment to the magnetic field direction and form nanoparticle chains, resulting in a fast switching speed. The only difference between the use of a parallel laser beam and that of a confocal laser beam is the energy distribution in the sample cell. Since more than one ultrahigh-order modes in the guiding layer can be excited by a confocal laser beam, the periodical structure of the electromagnetic energy distribution along the x direction is significantly disrupted. Experimental results in Fig. 8.47c show that the switching time is increased by three orders of magnitude simply, when energy distribution along the x direction is changed. These results show that the optical trapping effect is the main reason for the observed fast switching speed.

8.7.4 All-Optical Modulation [45, 46]

In this section, readers will find that the external magnetic field is not a prerequisite for the formation of magnetic nanoparticle aggregation. The experimental arrangement for all-optical modulation is shown in Fig. 8.52. The SMCW structure is firmly mounted on a computer-controlled $\theta/2\theta$ goniometer. A probe laser with a 650.00-nm wavelength passes subsequently through two apertures to be further collimated. The pump laser with an 860.00 nm wavelength is TE polarized and further collimated by a polarizer and two apertures, and its light intensity is tuned by an attenuator and real time measured with the help of a beam splitter and a power meter. These two laser beams are incident upon the SMCW structure at the same spot with two different incident angles. The incident angle of the pump laser is fixed at the bottom of one certain resonance dip, so the vast majority of the incident energy is coupled into the guiding layer and the light–matter interaction between the pump laser and the ferrofluid is strongest. Meanwhile, the incident angle of the probe laser locates at the middle point of the falling edge of another certain resonance dip, because at that point both the reflectivity and the GH shift exhibit a good linearity and a high sensitivity to the variation in the effective RI, which can be easily altered by the microstructure transition of ferrofluid in the SMCW structure. Since the intensity of the probe laser is much smaller compared with that of the pump one, the light–matter interaction induced by the probe laser can be neglected.

As shown in Fig. 8.53, the reflectivity of the probe laser reaches a peak when the pump laser power is about 10–50 MW. The dependence of the relative GH shift ($\Delta S = S_{\text{on}} - S_{\text{off}}$) on the control beam (pump laser) power is presented in Fig. 8.54a. With increasing the control beam power, ΔS drops rapidly to its minimum and then increases again slowly when the control beam power exceeds a critical power. Figure 8.54b shows one representative time evolution of ΔS upon switching on/off the control beam. These results reveal that the reflectivity and GH shift of the ferrofluid-sealed SMCW can be all optically modulated and there is a critical power for the control beam.

We suggest that the competition between the optical trapping effect and the Soret effect is most likely responsible for the critical power. When the control beam is absent, the magnetic nanoparticles are uniformly distributed in the carrier liquid.

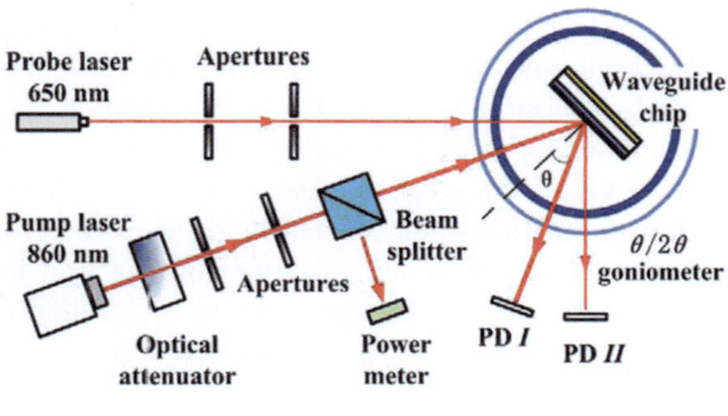

Fig. 8.52 Setup for all-optical modulation

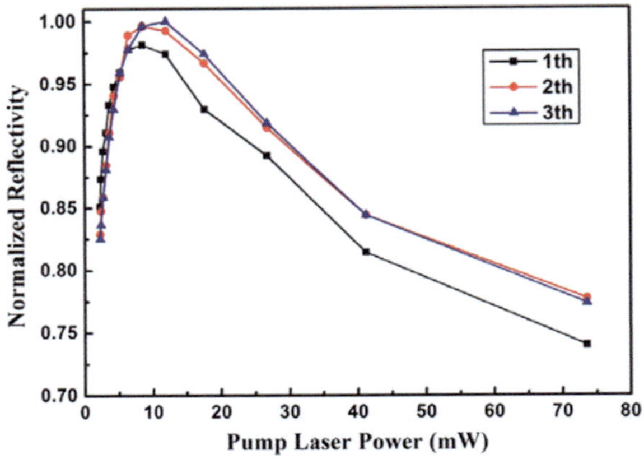

Fig. 8.53 Detected reflectivity (normalized to its maximum value) of the probe laser as a function of the pump laser power

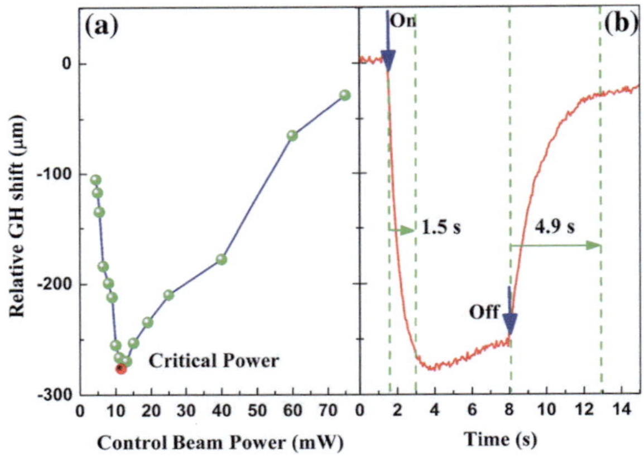

Fig. 8.54 **a** Relative GH shift of the signal beam as a function of the control beam power, **b** one representative time evolution of the relative GH shift upon switching on/off the control beam

Once the control beam is imposed, the interference of the ultrahigh-order mode between the two metal films will lead to a spatial periodic pattern of electromagnetic energy distribution perpendicular to metal surface. Therefore, the generated periodic energy pattern can push the magnetic nanoparticles to the high power density areas via the optical trapping effect, and the magnetic nanoparticles will aggregate together to form a periodic-like microstructure of ferrofluid. Since the aggregation process is reversible, the magnetic nanoparticle aggregations will dissolve again after the control beam is switched off. All these microstructure transitions of ferrofluid can lead to a variation in the resonance condition of the ultrahigh-order mode and finally give rise to a change in the reflectivity and the relative GH shift. Noted that the Soret effect is another factor contributing to the microstructure transitions of ferrofluid since the periodic energy pattern inevitably causes a thermal gradient, where the high power density areas are "hotter" and the low ones are "cooler." In contrast to the optical trapping effect, the Soret effect will cause the magnetic nanoparticles to escape from the "hotter" areas to the "cooler" areas. So the physical insight of the critical power in Figs. 8.53 and 8.54a can be understood as a consequence from the competition between the optical trapping effect and the Soret effect. As increasing the control beam power from 4.5 to 75.0 mW, if the control beam power is lower than the critical power, more magnetic nanoparticles will be concentrated into the high power density areas from the low ones due to the optical trapping effect; once the control beam power is larger than the critical power, the Soret effect will gradually become dominant and lead to a negative feedback, i.e., the magnetic nanoparticles will escape from the high power density areas to the low ones.

8.8 Self-assembly Concentric Circular Grating [47]

Optical micromanipulation, with its capability of moving or trapping the micro- or nanoobjects noninvasively by using light beam, has attracted significant research attention and motivated many applications in biology, colloidal dynamics, and particle sorting. Up to now, optical trapping of small objects such as dielectric spheres, metal nanoparticles, or even bioparticles has been demonstrated. The key operation of these optical trapping techniques is the enhancement of the gradient of light intensity, which determines the trapping efficiency. But, to achieve a large gradient, high NA lenses and a relatively high laser power are required. Another important issue, namely how to trap many particles simultaneously, is rarely reported.

The structure of SMCW is shown in Fig. 8.55. It consists of three glass slabs and two silver films: The bottom layer is a glass slab of 0.3 mm coated with a 300-nm-thick silver; the center layer is a 0.5-mm-thick glass slab with a rectangular cell of $10 \times 4\ \text{mm}^2$ working as the sample room; and the top layer is another thin glass slab (0.3 mm) coated with 35-nm-thick silver (Ag). All these components are assembled together by using the optical cement technology. As shown in Fig. 8.50,

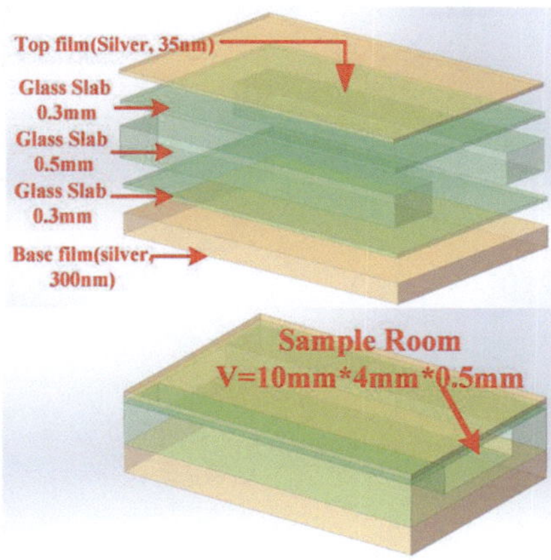

Fig. 8.55 Schematic diagram of the SMCW

the light energy is largely enhanced in the guiding layer when an ultrahigh-order mode is excited in the SMCW. The distribution of electromagnetic field of the excited ultrahigh-order mode will emerge as a set of concentric ring patterns (see Fig. 6.13c), which has been discussed in Sect. 6.5. The bright conical ring pattern implies strong energy in the guiding layer. If the analyte fluid containing nanoparticle is injected into the sample room of the SMCW, these nanoparticles can be trapped and stacked in the ring pattern field area. It should be mentioned that no micro- or nanostructure is preintroduced in our optical trapping method.

The experimental setup is shown in Fig. 8.56a. A pump laser beam comes from a 473-nm laser and incidents on a beam splitter. The reflected beam goes to a PD, which is used to monitor the output power. The transmission beam passes a hole in the projection screen and hits the SMCW, which is held on a rotation stage. By tuning the incident angle, an ultrahigh-order mode can be excited at some certain angle. The reflected conical concentric ring patterns are projected onto a screen and captured by a camera. Another beam coming from a He–Ne laser, whose incident power has been attenuated 80 %, is collimated and hit on the SMCW at a different angle. This beam is used as a probe beam to monitor the trapping of nanoparticles in the SMCW.

Nanoparticles were put into the solvent and mixed well. The prepared analyte is then injected into the sample room. When the incident laser beam is turned on and the stage is rotated to a resonance angle, the concentric ring pattern will be observed on the projection screen. Shortly after the excitement of an ultrahigh-order mode, the reflected beam of the probe laser shows a clear diffraction pattern (see Fig. 8.56a). Based on this phenomenon, we conjecture that nanoparticles are trapped in the area where the bright concentric ring pattern is formed. As the nanoparticles accumulate, they will form a refractive index distribution with a concentric ring pattern, which results in the diffraction pattern of the reflected probe

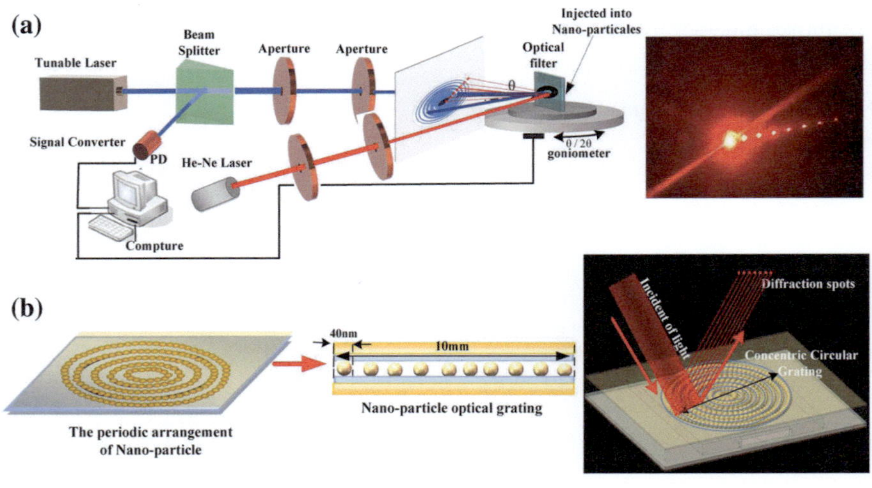

Fig. 8.56 Schematic diagram of the SMCW

beam. The trapping of the nanoparticles is stable during the experiment as long as the incident beam is held on. When the incident laser beam is turned off, nanoparticles become free and diffuse into the solution quickly under the effect of Brownian motion, while the diffraction pattern of the reflected probe beam disappears accordingly (see Fig. 8.56b).

References

1. H.G. Li, Z.Q. Cao, H.F. Lu, Q.S. Shen, Polarization-insensitive narrow band filter with a symmetrical metal-cladding optical waveguide. Chin. Phys. Lett. **23**, 643 (2006)
2. H.F. Lu, Z.Q. Cao, H.G. Li, Q.S. Shen, X.X. Deng, Polarization-independent and tunable comb filter based on a free-space coupling technique. Opt. Lett. **31**, 386 (2006)
3. C. Nylander, B. Liedberg, T. Lind, Gas detection by means of surface plasmon resonance. Sens. Actuators **3**, 79 (1982)
4. J. Homola, S.S. Yee, G. Gauglitz, Surface plasmon resonance sensors: review. Sens. Actuators, B **54**, 3 (1999)
5. X.D. Hoa, A.G. Kirk, M. Tabrizian, Towards integrated and sensitive surface plasmon resonance biosensors: a review of recent progress. Biosens. Bioelectron. **23**, 151 (2007)
6. T. Okamoto, M. Yamamoto, I. Yamaguchi, Optical waveguide absorption sensor using a single coupling prism. J. Opt. Soc. Am. A **17**, 1880 (2000)
7. R. Horvath, H.C. Pedersen, N. Skivesen, D. Selmeczi, N.B. Larsen, Monitoring of living cell attachment of spreading using reverse symmetry waveguide sensing. Appl. Phys. Lett. **86**, 071101 (2005)
8. R. Horvath, H.C. Pedersen, N.B. Larsen, Demonstration of reverse symmetry waveguide sensing in aqueous solutions. Appl. Phys. Lett. **81**, 2166 (2002)
9. G. Chen, Z.Q. Cao, J.H. Gu, Q.S. Shen, Oscillating wave sensors based on ultrahigh-order modes in symmetrical metal-clad optical waveguides. Appl. Phys. Lett. **89**, 081120 (2006)
10. J.H. Gu, G. Chen, Z.Q. Cao, Q.S. Shen, An intensity measurement refractometer based on a symmetric metal-clad waveguide structure. J. Phys. D Appl. Phys. **41**, 185105 (2008)
11. Y. Wang, M.Z. Huang, X.Y. Guan, Z.Q. Cao, F. Chen, X.P. Wang, Determination of trace chromium (VI) using a hollow-core metal-cladding optical waveguide sensor. Opt. Express **21**, 31130 (2013)
12. M. Born, E. Wolf, *Principles of Optics: Electromagnetic Theory of Propagation, Interference and Diffraction of Light* (Cambridge University Press, Cambridge, 1999)
13. Z. Han, L. Qi, G. Shen, W. Liu, Y. Chen, Determination of chromium(VI) by surface plasmon field-enhanced resonance light scattering. Anal. Chem. **79**, 5862 (2007)
14. J. Shi, Z.Q. Cao, J. Zhu, Q.S. Shen, Displacement measurement in real time using the attenuated total reflection technique. Appl. Phys. Lett. **84**, 3253 (2004)
15. F. Chen, Z.Q. Cao, Q.S. Shen, X.X. Deng, B.M. Duan, W. Yuan, M.H. Sang, S.Q. Wang, Nanoscale displacement measurement in a variable-air-gap optical waveguide. Appl. Phys. Lett. **88**, 161111 (2006)
16. F. Chen, Z.Q. Cao, Q.S. Shen, X.X. Deng, B.M. Duan, W. Yuan, M.H. Sang, S.Q. Wang, Picometer displacement sensing using the ultrahigh-order modes in a submillimeter scale optical waveguide. Opt. Express **13**, 10061 (2005)
17. F. Chen, Z.Q. Cao, Q.S. Shen, Y.J. Feng, Optical approach to angular displacement measurement based on attenuated total reflection. Appl. Opt. **44**, 5393 (2005)
18. S.Z. Zhang, S. Kiyono, Y. Uda, Nanoradian angle sensor and in situ self-calibration. Appl. Opt. **37**, 4154 (1998)
19. L. Chen, Z.Q. Cao, Q.S. Shen, X.X. Deng, F. Ou, Y.J. Feng, Wavelength sensing with subpicometer resolution using ultrahigh order modes. J. Lightw. Technol. **25**, 539 (2007)

20. Y. Wang, H.G. Li, Z.Q. Cao, T.Y. Yu, Q.S. Shen, Y. He, Oscillating wave sensor based on the Goos-Hänchen effect. Appl. Phys. Lett. **92**, 061117 (2008)
21. T.Y. Yu, H.G. Li, Z.Q. Cao, Y. Wang, Q.S. Shen, Y. He, Oscillating wave displacement sensor using the enhanced Goos-Hanchen effect in a symmetrical metal-cladding optical waveguide. Opt. Lett. **33**, 1001 (2008)
22. Y. Wang, X.G. Jiang, Q. Li, Y. Wang, Z.Q. Cao, High-resolution monitoring of wavelength shifts utilizing strong spatial dispersion of guided modes. Appl. Phys. Lett. **101**, 061106 (2012)
23. Y. Wang, Z.Q. Cao, T.Y. Yu, H.G. Li, Q.S. Shen, Enhancement of the superprism effect based on the strong dispersion effect of ultrahigh-order modes. Opt. Lett. **33**, 1276 (2008)
24. J. Hao, H.G. Li, C. Yin, Z.Q. Cao, 1.5 mm light beam shift arising from 14 pm variation of wavelength. J. Opt. Soc. Am. B **27**, 1305 (2010)
25. Y. Jiang, Z.Q. Cao, G. Chen, X.M. Dou, Y.L. Chen, Low voltage electro-optic polymer light modulator using attenuated total internal reflection. Opt. Laser Technol. **33**, 417 (2001)
26. Y. Jiang, Z.Q. Cao, X.M. Dou, Mechanism of a variable optical attenuator based on guided wave resonance. Chin. Phys. Lett. **18**, 1288 (2001)
27. X.X. Deng, X. Zheng, Z.Q. Cao, Q.S. Shen, H.G. Li, Fast speed electro-optic polymer variable optical attenuator based on cascaded attenuated-total-reflection technique. Appl. Phys. Lett. **90**, 151124 (2007)
28. Y. Wang, Z.Q. Cao, H.G. Li, Q.S. Shen, W. Yuan, P.P. Xiao, Tunable polarization beam splitting based on a symmetrical metal-cladding waveguide structure. Opt. Express **17**, 13309 (2009)
29. Y. Wang, Z.Q. Cao, H.G. Li, J. Hao, T.Y. Yu, Q.S. Shen, Electric control of spatial beam position based on the Goos-Hänchen shift. Appl. Phys. Lett. **93**, 091103 (2008)
30. H.C. Yang, I.J. Jang, H.E. Horng, J.M. Wu, Y.C. Chiou, C.Y. Hong, Pattern formation in microdrops of magnetic fluids. J. Magn. Magn. Mater. **201**, 313 (1999)
31. C.Y. Hong, Field-induced structural anisotropy in magnetic fluids. J. Appl. Phys. **85**, 5962 (1999)
32. C.Y. Hong, H.E. Horng, F.C. Kuo, S.Y. Yang, H.C. Yang, J.M. Wu, Evidence of multiple states of ordered structures and a phase transition in magnetic fluid films. Appl. Phys. Lett. **75**, 2196 (1999)
33. S. Taketomi, S. Ogawa, H. Miyajima, S. Chikazumi, Magnetic birefringence and dichroism in magnetic fluid. IEEE Trans. Magn. **4**, 384 (1989)
34. H.E. Horng, C.Y. Hong, S.L. Lee, C.H. Ho, S.Y. Yang, H.C. Yang, Magnetochromatics resulted from optical gratings of magnetic fluid films subjected to perpendicular magnetic fields. J. Appl. Phys. **88**, 5904 (2000)
35. X. Li, H. Ding, All-fiber magnetic-field sensor based on microfiber knot resonator and magnetic fluid. Opt. Lett. **37**, 5187 (2012)
36. T. Liu, X. Chen, Z. Di, J. Zhang, X. Li, J. Chen, Tunable magneto-optical wavelength filter of long-period fiber grating with magnetic fluids. Appl. Phys. Lett. **91**, 121116 (2007)
37. K.C. Neuman, S.M. Block, Optical trapping. Rev. Sci. Instrum. **75**, 2787 (2004)
38. Q. Dai, H. Deng, W. Zhao, J. Liu, L. Wu, S. Lan, A.V. Gopal, All-optical switching mediated by magnetic nanoparticles. Opt. Lett. **35**, 97 (2010)
39. N.V. Tabiryan, W. Luo, Soret feedback in thermal diffusion of suspensions. Phys. Rev. E **57**, 4431 (1998)
40. J.J. Sun, C. Yin, C.P. Zhu, X.P. Wang, W. Yuan, P.P. Xiao, X.F. Chen, Z.Q. Cao, Observation of magneto-optical effect in extremely dilute ferrofluids under weak magnetic field. J. Opt. Soc. Am. B **29**, 769 (2012)
41. H. Horng, C. Chen, K. Fang, S. Yang, J. Chieh, C. Hong, H. Yang, Tunable optical switch using magnetic fluids. Appl. Phys. Lett. **85**, 5592 (2004)
42. J. Li, X. Liu, Y. Lin, L. Bai, Q. Li, X. Chen, A. Wang, Field modulation of light transmission through ferrofluid film. Appl. Phys. Lett. **91**, 253108 (2007)

43. S. Yang, Y. Hsiao, Y. Huang, H. Horng, C. Hong, H. Yang, Retarded response of the optical transmittance through a magnetic fluid film under switching-on/off external magnetic fields. J. Magn. Magn. Mater. **281**, 48 (2004)
44. W. Yuan, C. Yin, P.P. Xiao, X.P. Wang, J.J. Sun, M.H. Sang, X.F. Chen, Z.Q. Cao, Microsecond-scale switching time of magnetic fluids due to the optical trapping effect in waveguide structure. Microfluid. Nanofluid. **11**, 781 (2011)
45. C. Yin, J.J. Sun, X.P. Wang, C.P. Zhu, Q.B. Han, Z.Y. Di, Z.Q. Cao, Modulated reflectivity via a symmetrical metal cladding ferrofluids core waveguide chip. EPL **100**, 44001 (2012)
46. X.P. Wang, C. Yin, J.J. Sun, H.G. Li, M.H. Sang, W. Yuan, Z.Q. Cao, M.Z. Huang, All-optically tunable Goos-Hänchen shift owing to the microstructure transition of ferrofluid in a symmetrical metal-cladding waveguide. Appl. Phys. Lett. **103**, 151113 (2013)
47. H.L. Dai, F. Ou, Z.Q. Cao, Y.X. Wang, H.G. Li, M.H. Sang, W. Yuan, F. Chen, X.F. Chen, Self-assembly concentric circular grating generated by the patterning trapping of nanoparticles in a micro fluidic chip (in preparation)

MIX
Papier aus verantwortungsvollen Quellen
Paper from responsible sources
FSC® C105338

If you have any concerns about our products,
you can contact us on
ProductSafety@springernature.com

In case Publisher is established outside the EU,
the EU authorized representative is:
Springer Nature Customer Service Center GmbH
Europaplatz 3, 69115 Heidelberg, Germany

Printed by Libri Plureos GmbH
in Hamburg, Germany